Getting Started with Raspberry Pi

Matt Richardson and Shawn Wallace

O'REILLY®

Beijing · Cambridge · Farnham · Köln · Sebastopol · Tokyo

Getting Started with Raspberry Pi

by Matt Richardson and Shawn Wallace

Published by O'Reilly Media, Inc., 1005 Gravenstein Highway North, Sebastopol, CA 95472.

O'Reilly books may be purchased for educational, business, or sales promotional use. Online editions are also available for most titles (*http://my.safaribooksonline.com*). For more information, contact our corporate/institutional sales department: 800-998-9938 or *corporate@oreilly.com*.

Editor: Brian Jepson
Production Editor: Christopher Hearse
Cover Designer: Randy Comer
Interior Designer: Ron Bilodeau and Edie Freedman
Illustrator: Marc de Vinck

December 2012:   First Edition

Revision History for the First Edition:

2012-12-07   First release

See *http://oreilly.com/catalog/errata.csp?isbn=9781449344214* for release details.

ISBN: 978-1-449-34421-4

LSI

Contents

Preface

It's easy to understand why people were skeptical of the Raspberry Pi when it was first announced. A credit card-sized computer for $35 seemed like a pipe dream. Which is why, when it started shipping, the Raspberry Pi created a frenzy of excitement.

Demand outstripped supply for months and the waitlists for these mini computers were very long. Besides the price, what is it about the Raspberry Pi that tests the patience of this hardware-hungry mass of people? Before we get into everything that makes the Raspberry Pi so great, let's talk about its intended audience.

Eben Upton and his colleagues at the University of Cambridge noticed that today's students applying to study computer science don't have the skills that they did in the 1990's. They attribute this to—among other factors—the "rise of the home PC and games console to replace the Amigas, BBC Micros, Spectrum ZX and Commodore 64 machines that people of an earlier generation learned to program on." Since the computer has become important for every member of the household, it may also discourage younger members from tinkering around and possibly putting such a critical tool out of commission for the family. But recently mobile phone and tablet processors have become less expensive while getting more powerful, clearing the path for the Raspberry Pi's leap into the world of ultra-cheap-yet-serviceable computer boards. As the founder of Linux, Linus Torvalds, said in an interview with BBC News, Raspberry Pi makes it possible to "afford failure."

What Can You Do With It?

One of the great things about the Raspberry Pi is that there's no single way to use it. Whether you just want to watch videos and surf the web, or you want to hack, learn, and make with the board, the Raspberry Pi is a flexible platform for fun, utility, and experimentation. Here are just a few of the different ways you can use a Raspberry Pi:

General purpose computing

It's important to remember that the Raspberry Pi is a computer and you can, in fact, use it as one. After you get it up and running in Chapter 1, you can choose to have it boot into a graphical desktop environment with a web browser, which is a lot of what we use computers for these days. Going beyond the web, you can install a wide variety of free software, such as the LibreOffice (*http://www.libreoffice.org/*) productivity suite for working with documents and spreadsheets when you don't have an Internet connection.

Learning to program

Since the Raspberry Pi is meant as an educational tool to encourage kids to experiment with computers, it comes preloaded with interpreters and compilers for many different programming languages. For the beginner, there's Scratch, a graphical programming language from MIT, which we cover in Chapter 5. If you're eager to jump into writing code, the Python programming language is a great way to get started and we cover the basics of it in Chapter 3. And you're not limited to only Scratch and Python. You can write programs for your Raspberry Pi in many different programming languages like C, Ruby, Java, and Perl.

Project platform

The Raspberry Pi differentiates itself from a regular computer not only in its price and size, but also because of its ability to integrate with electronics projects. Starting in Chapter 7, we'll show you to how to use the Raspberry Pi to control LEDs and AC devices and you'll learn how to read the state of buttons and switches.

Raspberry Pi for Makers

As makers, we have a lot of choices when it comes to platforms on which to build technology-based projects. Lately, microcontroller development boards like the Arduino have been a popular choice because they've become very easy to work with. But *System on a Chip* platforms like the Raspberry Pi are a lot different than traditional microcontrollers in many ways. In fact, the Raspberry Pi has more in common with your computer than it does with an Arduino.

This is not to say that a Raspberry Pi is better than a traditional microcontroller; it's just different. For instance, if you want to make a basic thermostat, you're probably better off using an Arduino Uno or similar microcontroller for purposes of simplicity. But if you want to be able to remotely access the thermostat via the web to change its settings and download temperature log files, you should consider using the Raspberry Pi.

Choosing between one or the other will depend on your project's requirements and in fact, you don't necessarily have to choose between the two. In Chapter 6, we'll show you how to use the Raspberry Pi to program the Arduino and get them communicating with each other.

As you read this book, you'll gain a better understanding of the strengths of the Raspberry Pi and how it can become another useful tool in the maker's toolbox.

But Wait... There's More!

There's so much you can do with the Raspberry Pi, we couldn't fit it all into one book. Here are a few other ways you can use it:

Media center
Since the Raspberry Pi has both HDMI and composite video outputs, it's easy to connect to televisions. It also has enough processing power to play full screen video in high definition. To leverage these capabilities, contributors to the free and open source media player, XBMC (*http://xbmc.org/*), have ported their project to the Raspberry Pi. XBMC can play many different media formats and its interface is designed with large buttons and text so that it can be easily controlled from the couch. XBMC makes the Raspberry Pi a fully customizable home entertainment center component.

"Bare metal" computer hacking
Most people who write computer programs write code that runs within an operating system, such as Windows, Mac OS, or—in the case of Raspberry Pi—Linux. But what if you could write code that runs directly on the processor without the need for an operating system? You could even write your own operating system from scratch if you were so inclined. The University of Cambridge's Computer Laboratory has published a free online course (*http://www.cl.cam.ac.uk/freshers/raspberrypi/tutorials/os/*) which walks you through the process of writing your own OS using assembly code.

Linux and Raspberry Pi

Your typical computer is running an operating system, such as Windows, OS X, or Linux. It's what starts up when you turn your computer on and it provides your applications access to hardware functions of your computer. For instance, if you're writing a application that accesses the Internet, you can use the operating system's functions to do so. You don't need to understand and write code for every single type of Ethernet or WiFi hardware out there.

Like any other computer, the Raspberry Pi also uses an operating system and the "stock" OS is a flavor of Linux called *Raspbian*. Linux is a great match for Raspberry Pi because it's free and open source. On one hand, it keeps the price of the platform low, and on the other, it makes it more hackable. And you're not limited to just Raspbian, as there are many different flavors, or *distributions*, of Linux that you can load onto the Raspberry Pi. There are even a few non-Linux OS options available out there. Throughout this book, we'll be using the standard Raspbian distribution that's available from Raspberry Pi's download page (*http://www.raspberrypi.org/downloads*).

If you're not familiar with Linux, don't worry, Chapter 2 will equip you with the fundamentals you'll need to know to get around.

What Others Have Done With It

When you have access to an exciting new technology, it can be tough deciding what to do with it. If you're not sure, there's no shortage of interesting and creative Raspberry Pi projects out there to get inspiration from. As editors for MAKE, we've seen a lot of fantastic uses of the Raspberry Pi come our way and we want to share some of our favorites.

Arcade Game Coffee Table (http://www.instructables.com/id/Coffee-Table-Pi/)
> Instructables user grahamgelding uploaded a step-by-step tutorial on how to make a coffee table that doubles as a classic arcade game emulator using the Raspberry Pi. To get the games running on the Pi, he used MAME (Multiple Arcade Machine Emulator), a free, open source software project which lets you run classic arcade games on modern computers. Within the table itself, he mounted a 24-inch LCD screen connected to the Raspberry Pi via HDMI, classic arcade buttons, and a joystick connected to the Pi's GPIO pins to be used as inputs.

RasPod (https://github.com/lionaneesh/RasPod)
> Aneesh Dogra, a teenager in India, was one of the runners up in Raspberry Pi Foundation's 2012 Summer Coding Contest. He created Raspod, a Raspberry Pi based web-controlled MP3 audio player. Built with Python and a web framework called Tornado, Raspod lets you remotely log into your Raspberry Pi to start and stop the music, change the volume, select songs, and make playlists. The music comes out of the Raspberry Pi's audio jack, so you can use it with a pair of computer speakers or you can connect it to a stereo system to enjoy the tunes.

Raspberry Pi Supercomputer (http://www.southampton.ac.uk/mediacentre/features/raspberry_pi_supercomputer.shtml)
> Many supercomputers are made of clusters of standard computers linked together and computational jobs are divided up among all the different processors. A group of computational engineers at the University of Southampton in the United Kingdom linked up 64 Raspberry Pis to create an inexpensive supercomputer. While it's nowhere near the

computational power of the top performing supercomputers of today, it demonstrates the principles behind engineering such systems. Best of all, the rack system used to hold all these Raspberry Pis was built with Lego bricks by the team leader's 6-year-old son.

If you do something interesting with your Raspberry Pi, we'd love to hear about it. You can submit your projects to the MAKE editorial team through our contribute form on Makezine.com (*http://blog.makezine.com/contribute/*).

Conventions Used in This Book

The following typographical conventions are used in this book:

Italic
> Indicates new terms, URLs, email addresses, filenames, and file extensions.

`Constant width`
> Used for program listings, as well as within paragraphs to refer to program elements such as variable or function names, databases, data types, environment variables, statements, and keywords.

`Constant width bold`
> Shows commands or other text that should be typed literally by the user.

`Constant width italic`
> Shows text that should be replaced with user-supplied values or by values determined by context.

 This icon signifies a tip, suggestion, or general note.

 This icon indicates a warning or caution.

Using Code Examples

This book is here to help you get your job done. In general, you may use the code in this book in your programs and documentation. You do not need to contact us for permission unless you're reproducing a significant portion of the code. For example, writing a program that uses several chunks of code from this book does not require permission. Selling or distributing a CD-ROM of examples from O'Reilly books does require permission. Answering a question by citing this book and quoting example code does not require permission. Incorporating a significant amount of example code from this book into your product's documentation does require permission.

We appreciate, but do not require, attribution. An attribution usually includes the title, author, publisher, and ISBN. For example: "*Getting Started With Raspberry Pi* by Matt Richardson and Shawn Wallace (O'Reilly). Copyright 2013, 978-1-4493-4421-4."

If you feel your use of code examples falls outside fair use or the permission given here, feel free to contact us at *permissions@oreilly.com*.

Safari® Books Online

 Safari Books Online is an on-demand digital library that lets you easily search over 7,500 technology and creative reference books and videos to find the answers you need quickly.

With a subscription, you can read any page and watch any video from our library online. Read books on your cell phone and mobile devices. Access new titles before they are available for print, get exclusive access to manuscripts in development, and post feedback for the authors. Copy and paste code samples, organize your favorites, download chapters, bookmark key sections, create notes, print out pages, and benefit from tons of other time-saving features.

O'Reilly Media has uploaded this book to the Safari Books Online service. To have full digital access to this book and others on similar topics from O'Reilly and other publishers, sign up for free at http://my.safaribooksonline.com (*http://my.safaribooksonline.com/?portal=oreilly*).

How to Contact Us

Please address comments and questions concerning this book to the publisher:

MAKE
1005 Gravenstein Highway North
Sebastopol, CA 95472
800-998-9938 (in the United States or Canada)
707-829-0515 (international or local)
707-829-0104 (fax)

MAKE unites, inspires, informs, and entertains a growing community of resourceful people who undertake amazing projects in their backyards, basements, and garages. MAKE celebrates your right to tweak, hack, and bend any technology to your will. The MAKE audience continues to be a growing culture and community that believes in bettering ourselves, our environment, our educational system—our entire world. This is much more than an audience, it's a worldwide movement that Make is leading—we call it the Maker Movement.

For more information about MAKE, visit us online:

MAKE magazine: *http://makezine.com/magazine/*
Maker Faire: *http://makerfaire.com*
Makezine.com: *http://makezine.com*
Maker Shed: *http://makershed.com/*

We have a web page for this book, where we list errata, examples, and any additional information. You can access this page at:

http://shop.oreilly.com/product/0636920023371.do

To comment or ask technical questions about this book, send email to:

bookquestions@oreilly.com

For more information about our books, courses, conferences, and news, see our website at *http://www.oreilly.com*.

Find us on Facebook: *http://facebook.com/oreilly*

Follow us on Twitter: *http://twitter.com/oreillymedia*

Watch us on YouTube: *http://www.youtube.com/oreillymedia*

Acknowledgements

We'd like to thank a few people who have provided their knowledge, support, advice, and feedback to *Getting Started with Raspberry Pi*:

Brian Jepson
Marc de Vinck
Eben Upton
Tom Igoe
Clay Shirky
John Schimmel
Phillip Torrone
Limor Fried
Kevin Townsend
Ali Sajjadi
Andrew Rossi

1/Getting Up and Running

A few words come up over and over when people talk about the Raspberry Pi: small, cheap, hackable, education-oriented. However, it would be a mistake to describe it as *plug and play*, even though it is easy enough to plug it into a TV set and get something to appear on the screen. This is not a consumer device, and depending on what you intend to do with your Raspberry Pi you'll need to make a number of decisions about peripherals and software when getting up and running.

Of course, the first step is to actually acquire a Raspberry Pi. Chances are you have one by now, but if not, the Raspberry Pi Foundation has arrangements with a few manufacturers from whom you can buy a Pi directly at the well-known $25-$35 price. They are:

Premier Farnell/Element 14 (http://www.element14.com/community/groups/raspberry-pi/)
> A British electronics distributor with many subsidiaries all over the world (such as Newark and MCM in the US).

RS Components (http://www.rs-components.com/raspberrypi)
> Another UK-based global electronics distributor (and parent of Allied Electronics in the US)

The low price of the Raspberry Pi is obviously an important part of the story. Enabling the general public to go directly to a distributor and order small quantities for the same price offered to resellers is an unusual arrangement. A lot of potential resellers were confounded by the original announcements of the price point; it was hard to see how there could be any profit margin. That's why you'll see resellers adding a slight markup to the $35 price (usually to $40 or so). Though the general public can still buy direct from the distributors above for the original price, the retailers and resellers often can fulfill orders faster. Both MAKE's own Maker Shed (*http://www.make rshed.com/category_s/227.htm*) as well as Adafruit (*http://www.adafruit.com/category/105*) are two companies who sell Raspberry Pis and accessories for a slight markup.

Enough microeconomic gossip; let's start by taking a closer look at the Raspberry Pi board.

A Tour of the Boards

Let's start with a quick tour of what you're looking at when you take it out of the box.

It's tempting to think of the Raspberry Pi as a microcontroller development board like Arduino, or as a laptop replacement. In fact it is more like the exposed innards of a mobile device, with lots of maker-friendly headers for the various ports and functions. Figure 1-1 shows all the parts of the board, as described below.

A. *The Processor*. At the heart of the Raspberry Pi is the same processor you would have found in the iPhone 3G and the Kindle 2, so you can think of the capabilities of the Raspberry Pi as comparable to those powerful little devices. This chip is a 32 bit, 700 MHz System on a Chip, which is built on the ARM11 architecture. ARM chips come in a variety of architectures with different *cores* configured to provide different capabilities at different price points. The Model B has 512MB of RAM and the Model A has 256 MB. (The first batch of Model Bs had only 256MB of RAM.)

B. *The Secure Digital (SD) Card slot*. You'll notice there's no hard drive on the Pi; everything is stored on an SD Card. One reason you'll want some sort of protective case sooner than later is that the solder joints on the SD socket may fail if the SD card is accidentally bent.

C. *The USB port*. On the Model B there are two USB 2.0 ports, but only one on the Model A. Some of the early Raspberry Pi boards were limited in the amount of current that they could provide. Some USB devices can draw up 500mA. The original Pi board supported 100mA or so, but the newer revisions are up to the full USB 2.0 spec. One way to check your board is to see if you have two *polyfuses* limiting the current (see Figure 1-2). In any case, it is probably not a good idea to charge your cell phone with the Pi. You can use a powered external hub if you have a peripheral that needs more power.

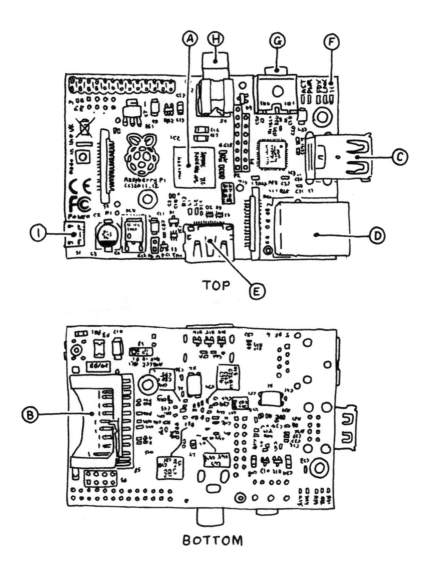

Figure 1-1. *A map of the hardware interface of the Raspberry Pi*

Figure 1-2. *Some of the older boards came equipped with polyfuses (left) to protect the USB hub. Some boards have the polyfuses replaced with jumpers (center), and the latest revision of the Model B removed them and uses the space for a mounting hole (right).*

D. *Ethernet port.* The model B has a standard RJ45 Ethernet port. The Model A does not, but can be connected to a wired network by a USB Ethernet adapter (the port on the Model B is actually an onboard USB to Ethernet adapter). WiFi connectivity via a USB dongle is another option.

E. *HDMI connector.* The HDMI port provides digital video and audio output. 14 different video resolutions are supported, and the HDMI signal can be converted to DVI (used by many monitors), composite (analog video signal usually carried over a yellow RCA connector), or SCART (a European standard for connecting audio-visual equipment) with external adapters.

F. *Status LEDs.* The Pi has five indicator LEDs that provide visual feedback (see Table 1-1).

Table 1-1. *The five status LEDs.*

ACT	Green	Lights when the SD card is accessed (marked OK on earlier boards)
PWR	Red	Hooked up to 3.3V power
FDX	Green	On if network adapter is full duplex
LNK	Green	Network activity light
100	Yellow	On if the network connection is 100Mbps (some early boards have a 10M misprint)

G. *Analog Audio output.* This is a standard 3.5mm mini analog audio jack, intended to drive high impedance loads (like amplified speakers). Headphones or unpowered speakers won't sound very good; in fact, as of this writing the quality of the analog output is much less than the HDMI audio output you'd get by connecting to a TV over HDMI. Some of this has to do with the audio driver software, which is still evolving.

H. *Composite video out*. This is a standard RCA-type jack that provides composite NTSC or PAL video signals. This video format is extremely low-resolution compared to HDMI. If you have a HDMI television or monitor, use it rather than a composite television.

I. *Power input*. On of the first things you'll realize is that there is no power switch on the Pi. This microUSB connector is used to supply power (this isn't an additional USB port; it's only for power). MicroUSB was selected because the connector is cheap USB power supplies are easy to find.

Figure 1-3 shows all of the power and input/output (*IO*) pins on the Raspberry Pi, which are explained next.

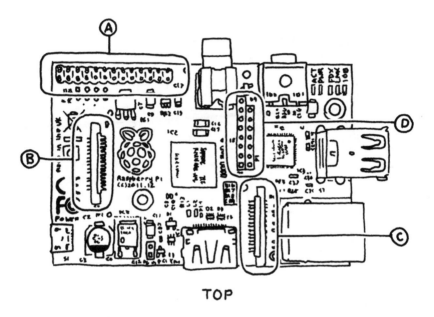

Figure 1-3. *The Pins and headers on the Raspberry Pi*

A. *General Purpose Input and Output (GPIO) and other pins*. Chapter 7 and Chapter 8 show how to use these pins to read buttons and switches and control actuators like LEDs, relays, or motors.

B. *The Display Serial Interface (DSI) connector*. This connector accepts a 15 pin flat ribbon cable that can be used to communicate with a LCD or OLED display screen.

C. *The Camera Serial Interface (CSI) connector*. This port allows a camera module to be connected directly to the board.

D. *P2 and P3 headers.* These two rows of headers are the JTAG testing headers for the Broadcom chip (P2) and the LAN9512 networking chip (P3). Because of the proprietary nature of the Broadcom chipset, these headers probably won't be of much use to you.

 In the Fall of 2012 the Raspberry Pi Foundation released a new revision of the board that included a few new hardware features, including two 2.5mm mounting holes and a header for a reset switch. There is also an unpopulated 2x4 header beneath the GPIO header that is intended for third-party clock and audio boards (to be mounted beneath the main board).

The Proper Peripherals

Now that you know where everything is on the board, you'll need to know a few things about the proper peripherals (some are shown in Figure 1-4) to use with the Pi. There are a bunch of prepackaged starter kits that have well-vetted parts lists; there are a few caveats and gotchas when fitting out your Raspberry Pi. There's a definitive list of supported peripherals (*http:// elinux.org/RPi_VerifiedPeripherals*) on the main wiki.

A. *A power supply.* This is the most important peripheral to get right; you should use a microUSB adapter that can provide 5V and at least 700mA of current (500mA for the Model A). A cell phone charger won't cut it, even if it has the correct connector. A typical cell phone charger only provides 400mA of current or less, but check the rating marked on the back. An underpowered Pi may still seem to work but will be flaky and may fail unpredictably.

 With the current version of the Pi board, it is possible to power the Pi from a USB hub that feeds power. However, there isn't much protection circuitry so it may not be the best idea to power it over the USB ports. This is especially true if you're going to be doing electronics prototyping where you may accidentally create shorts that may draw a lot of current.

B. *An SD Card.* You'll need at least 4GB, and it should be a Class 4 card. Class 4 cards are capable of transferring at least 4MB/sec. Some of the earlier Raspberry Pi boards had problems with Class 6 or higher cards, which are capable of faster speeds but are less stable. A microSD card in an adapter is perfectly usable as well.

Figure 1-4. *The basic peripherals: a microUSB power supply, cables, and SD card. You'll need at least a 4GB Class 4 SD Card (MicroSD cards with an adapter are ok to use as well). Generic SD Cards are notoriously variable in quality, so stick to a trusted model (see http://elinux.org/RPi_VerifiedPeriph erals#SD_cards).*

C. *An HDMI cable.* If you're connecting to a monitor you'll need this, or an appropriate adapter for a DVI monitor. You can also run the Pi headless, as described later in this chapter. HDMI cables can vary wildly in price. If you're just running a cable three to six feet to a monitor, there's no need to spend more than $3 USD on an HDMI cable. If you are running long lengths, you should definitely research the higher quality cables and avoid the cheap generics.

D. *Ethernet cable.* Your home may not have as many wired Ethernet jacks as it did five years ago. Since everything is wireless these days, you might find the wired port to be a bit of a hurdle; see the section "Running Headless" (page 31) for some alternatives to plugging the Ethernet directly into the wall or a hub.

If you want to do a lot more with your Raspberry Pi there are a few other peripherals and add ons that you'll want, which we'll talk about in Chapter 5. You may also want to consider some of the following add ons (see *http://elinux.org/RPi_VerifiedPeripherals* for a list of peripherals that are known to work):

A Powered USB Hub
A USB 2.0 hub is recommended.

Heatsink
A heatsink is a small piece of metal, usually with fins to create a lot of surface area to dissipate heat efficiently. Heatsinks can be attached to chips that get hot. The Pi's chipset was designed for mobile applications, so a heatsink isn't necessary most of the time. However, as we'll see later there are cases where you may want to run the Pi at higher speeds, or crunch numbers over an extended period and the chip may heat up a bit. Some people have reported that the network chip can get warm as well.

Real Time Clock
You may want to add a Real Time Clock chip (like the DS1307) for logging or keeping time when offline.

Camera module
An official 5 megapixel Raspberry Pi camera module will be available in early 2013. Until then you can use a USB web cam (see Chapter 9 for a complete example).

LCD display
Many LCDs can be used via a few connections on the GPIO header. LCDs that use the DSI interface will be available in 2013.

WiFi USB dongle
Many WiFi USB dongles work with the Pi; look for one that doesn't draw too much power.

Laptop dock
Several people have modified laptop docks intended for cell phones (like the Atrix lapdock) to work as a display/base for the Raspberry Pi.

The Case

You'll quickly find that you'll want a case for your Raspberry Pi. The stiff cables on all sides make it hard to keep flat, and some of the components like the SD card slot can be mechanically damaged even through normal use.

The Pi contains six layers of conductive traces connecting various components, unlike a lot of simple microcontroller PCBs that just have traces on the top and the bottom. There are four layers of thin traces sandwiched in between the top and bottom; if the board gets flexed too much you can break some of those traces in a manner that is impossible to debug. The solution: get a case.

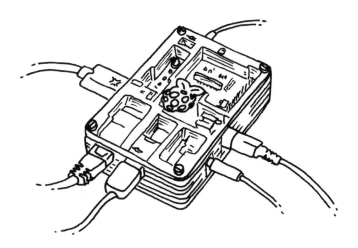

Figure 1-5. *The colorful Pibow case*

There are a bunch of pre-made cases available, but there are also a lot of case designs available to download and fabricate on a laser cutter or 3D printer. In general, avoid tabbed cases where brittle acrylic is used at right angles. The layered acrylic of the Pibow (*http://pibow.com/*) (Figure 1-5) is a colorful option.

It should probably go without saying, but it's one of those obvious mistakes you can make sometimes: make sure you don't put your Raspberry Pi on a conductive surface. Flip over the board and look at the bottom; there are a lot of components there and a lot of solder joints that can be easily shorted. Another reason why it's important to case your Pi! === Choose Your Distribution

The Raspberry Pi runs Linux for an operating system. Linux is technically just the kernel, and an operating system is much more than that; the total collection of drivers, services, and applications makes the OS. A variety of flavors or distributions of Linux the OS have evolved over the years. Some of the most common on desktop computers are Ubuntu, Debian, Fedora, and Arch. Each have their own communities of users and are tuned for particular applications.

Because the Pi is based on a mobile device chipset, it has different software requirements than a desktop computer. The Broadcom processor has some proprietary features that require special "binary blob" device drivers and code that won't be included in any standard Linux distribution. And, while

Figure 1-6. *Raspberry Pi + Debian = Raspbian.*

most desktop computers have gigabytes of RAM and hundreds of gigabytes of storage, the Pi is more limited in both regards. Special Linux distributions that target the Pi have been developed. Some of the more established distributions are:

Raspbian (http://raspbian.org)
> The "officially recommended" official distribution from the Foundation, based on Debian (see Figure 1-6). Note that raspbian.org is a community site, not operated by the Foundation. If you're looking for the official distribution, visit the downloads (*http://raspberrypi.org*) page at raspberrypi.org.

Adafruit Raspberry Pi Educational Linux (Occidentalis) (http://learn.adafruit.com/adafruit-raspberry-pi-educational-linux-distro)
> This is Adafruit's Raspbian-based distribution that includes tools and drivers useful for teaching electronics.

Arch Linux (http://www.archlinux.org/)
> Arch Linux specifically targets ARM-based computers, so they supported the Pi very early on.

Xbian (http://xbian.org/)
> This is a distribution based on Raspbian for users who want to use the Raspberry Pi as a media center (see also OpenELEC (*http://openelec.tv/*) and Raspbmc (*http://www.raspbmc.com*)).

QtonPi (http://qt-project.org/wiki/Qt-RaspberryPi)
> A distribution based on the Qt 5 framework.

In this book we will concentrate on the official Raspbian distribution.

Flash the SD Card

Many vendors sell SD cards with the operating system pre-installed; for some people this may be the best way to get started. Even if it isn't the latest release you can easily upgrade once you get the Pi booted up and on the Internet.

Raspbian also has a network installer (*http://www.raspbian.org/Raspbia nInstaller*). To use this tool, you need to put the installer files on a SD Card (formatted as FAT32, which is typical for these cards) and then boot up the Pi with the card inserted. The catch is that you'll need to be connected to the Internet for this to work.

The first thing you'll need to do is download Raspbian from here the downloads page at raspberry pi.org (*http://www.raspberrypi.org/downloads*). The operating system is distributed as a disk image, which is a bit-for-bit representation of how the data should be written to the SD card.

Note that you can't just drag the disk image onto the SD Card; you'll need to make a bit-for-bit copy of the image. You'll need a card writer and a disk image utility; any inexpensive card writer will do. The instructions vary depending on the OS you're running. Unzip the image file (you should end up with a *.img* file), then follow the appropriate directions, as described in Appendix A.

Faster Downloads With BitTorrent

You'll see a note on the download site about downloading a torrent file for the most efficient way of downloading Raspbian. The torrent file is a decentralized way of distributing files; it can be much faster because you'll be pulling bits of the download from many other torrent clients rather than a single central server. You'll need a BitTorrent client if you choose this route.

Some popular BitTorrent clients are:

- Vuze (*http://www.vuze.com/*): Integrated torrent search and download.
- Miro (*http://www.getmiro.com/*): open source music and video player that also handles torrents.
- MLDonkey (*http://mldonkey.sourceforge.net/Main_Page*): Windows and Linux-only filesharing tool.
- Transmission (*http://www.transmissionbt.com/*): Lightweight Mac and Linux-only client, also used in embedded systems.

The operating system is distributed as a disk image, which is a bit-for-bit representation of how the data should be written to the SD card. Because it is an image, creating a bootable card is not as simple a process as dragging files on or off the card on your desktop. You'll need a card writer and a disk image utility; any inexpensive card writer will do. The instructions vary depending on the OS you're running. Unzip the image file (you should end up with a *.img* file), then follow the appropriate directions below.

Booting Up

Follow these steps to book up your Raspberry Pi for the first time:

1. Plug the SD card into the socket.
2. Plug in a USB keyboard and mouse. On the Model A, plug them into a powered hub, then plug the hub into the Pi.
3. Plug the HDMI output into your TV or monitor. Make sure your monitor is on.
4. Plug in the power supply. In general, try to make sure everything else is hooked up before connecting the power.

Getting Online

You've got a few different ways to connect to the Internet. If you've got easy access to a router, switch (or Ethernet jack connected to a router), just plug in using a standard Ethernet cable. If you have a WiFi USB dongle, you can connect wirelessly; there's an icon on the Desktop to set up your wireless connection. Not all dongles will work; check the verified peripherals list (*http://elinux.org/RPi_VerifiedPeripherals*) for a supported one.

If you've got a laptop nearby, or if you're running the Pi in a headless configuration, you can share the WiFi on your laptop with the Pi (Figure 1-7). It is super simple on the Mac: just enable Internet Sharing In your Sharing settings, then use an Ethernet cable to connect the Pi and your Mac. In Windows, enable "Allow other network users to connect through this computer's Internet connection" in your Internet Connection Sharing properties. The Pi should automatically get an IP address when connected and be online.

You will probably need a cross-over cable for a Windows-based PC, but you can use any Ethernet cable on Apple hardware as it will autodetect the type of cable.

If all goes well, you should see a bunch of startup log entries appearing on your screen. If things don't go well, skip ahead to the troubleshooting section at the end of this chapter. These log messages show all of the processes that are launching as you boot up the Pi. You'll see the network interface be initialized, and you'll see all of your USB peripherals being recognized and logged. You can see these log messages after you log in by typing dmesg on the command line.

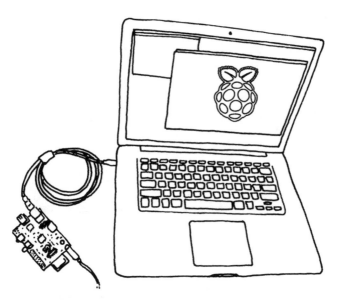

Figure 1-7. *A handy trick is to share your laptop's WiFi connection with the Pi. You can also run the Pi headless (see "Running Headless" (page 31)), which is convenient if you're using your Raspberry Pi on the run.*

The very first time you boot up you'll be presented with a few the raspi-config tool (see Figure 1-8). There are a few key settings you'll need to tweak here; chances are good that your Raspberry Pi won't work exactly the way you want right out of the box. If you need to get back to this configuration tool at any time by typing the following at the command line:

```
sudo raspi-config
```

Configuring Your Pi

Next we'll walk through the steps and show you which configuration options are essential and which you can come back to if you need them. When setting options in the tool, use the up and down arrows to move around the list, the space bar to select something, and tab to change fields or move the cursor to the buttons at the bottom of the window. Let's go in the order of the menu options in the configuration tool:

Expand rootfs
> You should always choose this option; this will enlarge the filesystem to let you use the whole SD card.

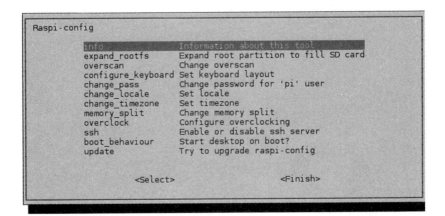

Figure 1-8. *The Raspi-config tool menu*

Overscan

Leave the overscan option disabled at first. If you have a high definition monitor you may find that text runs off the side of the screen. To fix this, enable the overscan and change the values to fit the image to the screen. The values indicate the amount of overscan so the display software can correct; use positive values if the image goes off the screen, negative if there are black borders around the edge of the display.

Keyboard

The default keyboard settings are for a generic keyboard in a UK-style layout. If you want they keys to do what they're labeled to do, you'll definitely want to select a keyboard type and mapping that corresponds to your setup. Luckily the keyboard list is very robust. Note that your locale settings can affect your keyboard settings as well.

Password

It's a good idea to change the default password from *raspberry* to something a little stronger.

Change Locale

If you're outside the UK you should change your locale to reflect your language and character encoding preferences. The default setting is for UK English with a standard UTF-8 character encoding (en_GB.UTF-8). Select en_US.UTF-8 if you're in the US.

Change timezone

You'll probably want to set this.

Memory split

This option allows you to change the amount of memory used by the CPU and the GPU. Leave the default split for now.

Overclock

You now have the option of running the processor at speeds higher than 700MHz with this option. For your first time booting, leave the default settings or try *Medium* or *Modest*. You may want to return to this later (Turbo mode can run at 1000MHz).

SSH

This option turns on the Secure Shell (ssh) server, which will allow you to login to the Raspberry Pi remotely over a network. This is really handy, so you should turn it on.

Desktop Behavior

This option lets you boot straight to the graphical desktop environment and is set to Yes by default. If you select No, you'll get the command line when you boot up and you'll have to login and start the graphical interface manually like this:

```
raspberrypi login: pi
Password: raspberry
pi@raspberrypi ~ $ startx
```

When you're in the graphical desktop, your command prompt will disappear. You can open a terminal program to get a command prompt while you're in the graphical desktop. Click the desktop menu in the lower left, then choose Accessories→LXTerminal.

Update

Finally, if you're connected to the Internet you'll be able to update the conifg utility with this option. Don't update the OS on your first time around; you'll see other ways to do this in Chapter 2.

When you're done, select Finish and you'll be dumped back to the command line. Type:

```
pi@raspberrypi ~ $ sudo reboot
```

and your Pi will reboot with your new settings. If all goes well (and if you chose the option to boot straight to the graphical desktop environment) you should see the Openbox window manager running on the Lightweight X11 Desktop Environment (LXDE). You're off and running!

Shutting Down

There's no power button on the Raspberry Pi (although there is a header for a reset switch on newer boards). The proper way to shutdown is through the Logout menu on the graphical desktop; select Shutdown to halt the system.

You can also shut down from the command line by typing:

```
pi@raspberrypi ~ $ sudo shutdown -h now
```

Be sure to do a clean shutdown (and don't just pull the plug). In some cases you can corrupt the SD card if you turn off power without halting the system.

Troubleshooting

If things aren't working the way you think they should, there are a few common mistakes and missed steps. Be sure to check all of the following suggestions:

- Is the SD card in the slot, and is it making a good connection? Are you using the correct type of SD Card?
- Was the disk image written correctly to the card? Try copying it again with another card reader.
- Is the write protect enabled on SD card? This is a little switch on the side that can easily get toggled the wrong way.
- Check the integrity of your original disk image. You can do this by running a Secure Hash Algorithm (SHA) checksum utility on the disk image and comparing the result to the 40 character hash published on the download page.
- Is the Pi restarting or having intermittent problems? Check your power supply; an underpowered board may seem to work but act flaky.
- Do you get a kernel panic on startup? A kernel panic is the equivalent of Windows' Blue Screen of Death; it's most often caused by a problem with a device on the USB hub. Try unplugging USB devices and restarting.

If that all fails, head over to the troubleshooting page on the Raspberry Hub wiki (*http://elinux.org/R-Pi_Troubleshooting*) for solutions to all sorts of problems people have had.

 Which Board Do You Have?

If you're asking for help in an email or on a forum, it can be helpful to the helper if you know exactly what version of the operating system and which board you're using. To find out the OS version, open LXTerminal and type:

```
cat /proc/version
```

To find your board version, type:

```
cat /proc/cpuinfo
```

Going Further

The Raspberry Pi Hub (http://elinux.org/RPi_Hub)
> Hosted by elinux.org, this is a massive Wiki of information on the Pi's hardware and configuration.

List of Verified Peripherals (http://elinux.org/RPi_VerifiedPeripherals)
> The definitive list of peripherals known to work with the Raspberry Pi.

2/Getting Around Linux on the Raspberry Pi

If you're going to get the most out of your Raspberry Pi, you'll need to learn a little Linux. The goal of this chapter is to present a whirlwind tour of the operating system and give you enough context and commands to get around the file system, install packages from the command line or GUI, and point out the most important tools you'll need day to day.

Raspbian comes with the Lightweight X11 Desktop Environment (LXDE) graphical desktop environment installed. This is a trimmed-down desktop environment for the X Window System that has been powering the GUIs of Unix and Linux computers since the 80s. Some of the tools you see on the Desktop and in the menu are bundled with LXDE (the Leafpad text editor and the LXTerminal shell, for instance).

Running on top of LXDE is Openbox, a *window manager* that handles the look and feel of windows and menus. If you want to tweak the appearance of your desktop, go to the Openbox configuration tools (click the desktop menu in the lower left, then choose Other→Openbox Configuration Manager). Unlike OS X or Windows, it is relatively easy to completely customize your desktop environment or install alternate window mangers. Some of the other distributions for Raspberry Pi have different environments tuned for applications like set top media boxes, phone systems or network firewalls. See *http:// elinux.org/RPi_Distributions* for a list.

The File Manager
> If you prefer not to move files around using the command line (more on that in a moment), select the File Manager from the Accessories menu. You'll be able to browse the filesystem using icons and folders the way you're probably used to doing.

Figure 2-1. *The graphical desktop.*

The Web Browser

The default web browser is Midori (*http://twotoasts.de/index.php/midori/*), designed to work well with limited resources. It's easy to forget how much work web browsers do these days. Because Raspbian is designed to be a very lightweight OS distribution, there are a number of features you may expect in a web browser that are not available. For example Flash and the Java plugin is not installed (so no YouTube), and Midori does not support HTML5 video. Later we'll look at how to install new software (like Java). Look for tools and menu items in the pulldown menu in the upper right corner of the window (see Figure 2-2). There are a couple of other browser options, notably NetSurf (*http://www.netsurf-browser.org/*) and Dillo (*http://www.dillo.org/*).

Video and Audio

Multimedia playback is handled by `omxplayer`, which is a bit experimental as of this writing. It is only available as a command line utility. Omxplayer is specially designed to work with the Graphics Processing Unit (GPU) on the processor; other free software like VLC and mPlayer won't work well with the GPU.

To keep the price down, certain video licenses were not included with the Raspberry Pi. If you want to watch recorded TV and DVDs encoded in the MPEG-2 format (or Microsoft's VC-1 format), you'll need to purchase an license key from the Foundation's online shop (*http://raspberrypi.com/*). A license for H.264 (MPEG-4) decoding and encoding is included with the Raspberry Pi.

Figure 2-2. *The pulldown menu in the web browser.*

Text Editor

Leafpad (*http://wiki.lxde.org/en/Leafpad*) is the default text editor, which is available under the main menu. You've also got Nano (*http://www.nano-editor.org/*), which is an easy-to-learn bare bones text editor. Traditional Unix text editors like vim or emacs are not installed by default, but can be easily added (see "Installing New Software" (page 32)).

Copy and Paste

Copy and paste functions work between applications pretty well, although you may find some oddball programs that aren't consistent. If you have a middle button on your mouse you can select text by highlighting it as you normally would (click and drag with the left mouse button) and paste it by pressing the middle button while you have the mouse cursor over the destination window.

The Shell

A lot of tasks are going to require you to get to the command line and run commands there. The LXTerminal program provides access to the command line or shell. The default shell on Raspbian is the Bourne Again Shell (bash (*http://www.gnu.org/software/bash/manual/bash*

ref.html)), which is very common on Linux systems. There's also an alternative called dash (*http://en.wikipedia.org/wiki/Debian_Almquist_shell*). You can change shells via the program menu, or with the chsh command.

Using the Command Line

If it helps, you can think of using the command line as playing a text adventure game, but with the files and the filesystem in place of Grues and mazes of twisty passages. If that metaphor doesn't help you, don't worry; all the commands and concepts in this section are standard Linux and are valuable to learn.

Before you start, open up the LXTerminal program (Figure 2-3). There are two tricks that make life much easier in the shell: *autocomplete* and *command history*. Often you will only need to type the first few characters of a command or filename, then hit tab. The shell will attempt to autocomplete the string based on the files in the current directory or programs in commonly used directories (the shell will search for executable programs in places like /bin or /usr/bin/). If you hit the up arrow on the command line you'll be able to step back through your command history, which is useful if you mistyped a character in a long string of commands.

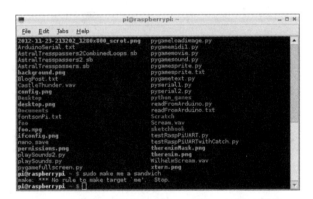

Figure 2-3. *LXTerminal gives you access to the command line (or shell).*

Files and the Filesystem

Table 2-1 shows some of the important directories in the filesystem. Most of these follow the Linux standard where files should go; a couple are specific to the Raspberry Pi. The /sys directory is where you can access all of the hardware on the Raspberry Pi.

Table 2-1. *Some of the most important directories in the Raspbian filesystem.*

Directory	Description
/	
/bin	Programs and commands that all users can run
/boot	All the files needed at boot time
/dev	Special files that represent the devices on your system
/etc	Configuration files
/etc/init.d	Scripts to start up services
/etc/X11	X11 configuration files
/home	User home directories
/home/pi	Home directory for pi user
/lib	Kernel modules/drivers
/media	Mount points for removable media
/proc	A virtual directory with information about running processes and the OS
/sbin	Programs for system maintenance
/sys	A special directory on the Raspberry Pi that represents the hardware devices
/tmp	Space for programs to create temporary files
/usr	Programs and data usable by all users
/usr/bin	Most of the programs in the operating system reside here
/usr/games	Yes, games
/usr/lib	Libraries to support common programs
/usr/local	Software that may be specific to this machine goes here
/usr/sbin	More system administration programs
/usr/share	Things that are shared between applications like icons or fonts
/usr/src	Linux is open source; here's the source!
/var	System logs and spool files
/var/backups	Backup copies of all the most vital system files
/var/cache	Any program that caches data (like apt-get or a web browser) stores it here.
/var/log	All of the system logs and individual service logs
/var/mail	All user email is stored here, if you're set up to handle email
/var/spool	Data waiting to be processed (e.g. incoming email, print jobs)

You'll see your current directory displayed before the command prompt. In Linux your home directory has a shorthand notation: the tilde (~). When you open the LXTerminal you'll be dropped into your home directory and your prompt will look like this:

```
pi@raspberrypi ~ $
```

Here's an explanation of that prompt:

pi@❶raspberrypi❷ ~❸ $❹

❶   Your username, pi, followed by the at (@) symbol.

❷   The name of your computer (raspberrypi is the default host name).

❸   The *current working directory* of the shell. You always start out in your home directory (~).

❹   This is the *shell prompt*. Any text you type will appear to the right of it. Press Enter or Return to execute each command you type.

Use the cd (change directory) command to move around the filesystem. The following two commands have the same effect (changing to the home directory) for the pi user:

```
cd /home/pi/
cd ~
```

If the directory path starts with a forward slash it will be interpreted as an absolute path to the directory. Otherwise the directory will be considered relative to the current working directory. You can also use . and .. to refer to the current directory and the current directory's parent. For example, to move up to the top of the filesystem:

```
pi@raspberrypi ~ $ cd ..
pi@raspberrypi /home $ cd ..
```

You could also get there with the absolute path /:

```
pi@raspberrypi ~ $ cd /
```

Once you've changed to a directory, use the ls command to list the files there.

```
pi@raspberrypi / $ ls
bin    dev   home  lost+found  mnt   proc  run   selinux  sys   usr
boot   etc   lib   media       opt   root  sbin  srv      tmp   var
```

Most commands have additional parameters or *switches* that can be used to turn on different behaviors. For example, the -l switch will produce a more detailed listing, showing file sizes, dates and permissions:

```
pi@raspberrypi ~ $ ls -l
total 8
drwxr-xr-x 2 pi pi 4096 Oct 12 14:26 Desktop
drwxrwxr-x 2 pi pi 4096 Jul 20 14:07 python_games
```

The -a switch will list all files, including invisible ones:

```
pi@raspberrypi ~ $ ls -la
total 80
drwxr-xr-x 11 pi   pi   4096 Oct 12 14:26 .
drwxr-xr-x  3 root root 4096 Sep 18 07:48 ..
-rw-------  1 pi   pi     25 Sep 18 09:22 .bash_history
-rw-r--r--  1 pi   pi    220 Sep 18 07:48 .bash_logout
-rw-r--r--  1 pi   pi   3243 Sep 18 07:48 .bashrc
drwxr-xr-x  6 pi   pi   4096 Sep 19 01:19 .cache
drwxr-xr-x  9 pi   pi   4096 Oct 12 12:57 .config
drwx------  3 pi   pi   4096 Sep 18 09:24 .dbus
drwxr-xr-x  2 pi   pi   4096 Oct 12 14:26 Desktop
-rw-r--r--  1 pi   pi     36 Sep 18 09:35 .dmrc
drwx------  2 pi   pi   4096 Sep 18 09:24 .gvfs
drwxr-xr-x  2 pi   pi   4096 Oct 12 12:53 .idlerc
-rw-------  1 pi   pi     35 Sep 18 12:11 .lesshst
drwx------  3 pi   pi   4096 Sep 19 01:19 .local
-rw-r--r--  1 pi   pi    675 Sep 18 07:48 .profile
drwxrwxr-x  2 pi   pi   4096 Jul 20 14:07 python_games
drwx------  4 pi   pi   4096 Oct 12 12:57 .thumbnails
-rw-------  1 pi   pi     56 Sep 18 09:35 .Xauthority
-rw-------  1 pi   pi    300 Oct 12 12:57 .xsession-errors
-rw-------  1 pi   pi   1391 Sep 18 09:35 .xsession-errors.old
```

Use the mv command to rename a file. The **touch** command can be used to create an empty dummy file:

```
pi@raspberrypi ~ $ touch foo
pi@raspberrypi ~ $ ls
foo    Desktop    python_games
pi@raspberrypi ~ $ mv foo baz
pi@raspberrypi ~ $ ls
baz    Desktop    python_games
```

Remove a file with rm. To remove a directory you can use rmdir if the directory if empty, or rm -r if it isn't. The -r is a parameter sent to the rm command that indicates it should recursively delete everything in the directory.

If you want to find out all the parameters for a particular command, use the man command (or you can often use the --help option):

```
pi@raspberrypi ~ $ man curl
pi@raspberrypi ~ $ rm --help
```

To create a new directory, use mkdir. To bundle all of the files in a directory into a single file, use the **tar** command, originally created for tape archives. You'll find a lot of bundles of files or source code are distributed as tar files, and they're usually also compressed using the gzip command. Try this:

```
pi@raspberrypi ~ $ mkdir myDir
pi@raspberrypi ~ $ cd myDir
```

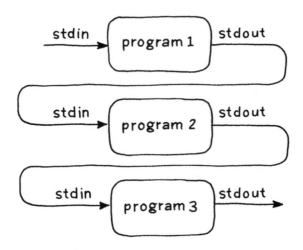

Figure 2-4. *Pipes are a way of chaining smaller programs together to accomplish bigger tasks.*

```
pi@raspberrypi ~ $ touch foo bar baz
pi@raspberrypi ~ $ cd ..
pi@raspberrypi ~ $ tar -cf myDir.tar myDir
pi@raspberrypi ~ $ gzip myDir.tar
```

You'll now have a *.tar.gz* archive of that directory that can be distributed via email or Internet.

More Linux Commands

One of the reasons that Linux (and Unix) are so successful is that the main design goal was to build a very complicated system out of small, simple modular parts that can be chained together. You'll need to know a little bit about two pieces of this puzzle: *pipes* and *redirection*.

Pipes are simply a way of chaining two programs together, so the output of one can serve as the input to another. All Linux programs can read data from *standard input* (often referred to as *stdin*), write data to *standard output* (*stdout*), and throw error messages to *standard error* (*stderr*). A pipe lets you hook up stdout from one program to stdin of another (Figure 2-4). Use the | operator, as in this example:

```
pi@raspberrypi ~ $ ls -la | less
```

(Press q to exit the less program.)

Now (for something a little more out there) try:

```
pi@raspberrypi ~ $ sudo cat /boot/kernel.img | aplay
```

You may want to turn the volume down a bit first; this command reads the kernel image and spits all of the 1s and 0s at the audio player. That's what your kernel sounds like!

In some of the examples later in the book we'll also be using *redirection*, where a command is executed and the stdout output can be sent to a file. As you'll see later, many things in Linux are treated as ordinary files (such as the Pi's general purpose input and output pins), so redirection can be quite handy. To redirect output from a program use the > operator:

```
pi@raspberrypi ~ $ ls > directoryListing.txt
```

Special Control Keys

In addition to the keys for autocomplete (tab) and command history (up arrow) previously mentioned, there are a few other special control keys you'll need in the shell. Here are a few:

Control-C
Kills the currently running program. May not work with some interactive programs such as text editors.

Control-D
Exits the shell. You must type this at the command prompt by itself (don't type anything after the $ before hitting Control-D).

Control-A
Moves the cursor to the beginning of the line.

Control-E
Moves the cursor to the end of the line.

There are others but these are the core keyboard shortcuts you'll use every day.

Sometimes you'll want to display the contents of a file on the screen. If it's a text file and you want to read it one screen at a time, use less:

```
pi@raspberrypi ~ $ ls > flob.txt
pi@raspberrypi ~ $ less flob.txt
```

If you want to just dump the entire contents of a file to standard output, use cat (short for concatenate). This can be handy when you want to feed a file into another program or redirect it somewhere.

For example, this is the equivalent of copying one file to another with a new name (the second line concatenates the two files first):

```
pi@raspberrypi ~ $ ls > wibble.txt
pi@raspberrypi ~ $ cat wibble.txt > wobble.txt
pi@raspberrypi ~ $ cat wibble.txt wobble.txt > wubble.txt
```

To look at just the last few lines of a file (such as the most recent entry in a log file), use `tail` (to see the beginning, use `head`). If you are searching for a string in one or more files, use the venerable program `grep`:

```
pi@raspberrypi ~ $ grep Puzzle */*
```

Grep is a powerful tool because of the rich language of *regular expressions* that was developed for it. Regular expressions can be a bit difficult to read, and may be a major factor in whatever reputation Linux has for being opaque to newcomers.

Processes

Every program on the Pi runs as a separate process; at any particular point in time you'll have dozens of processes running. When you first boot up, about 75 processes will start, each one handling a different task or service. To see all these processes, run the `top` program, which will also display CPU and memory usage. Top will show you the processes using the most resources; use the `ps` command to list all the processes and their id numbers. Try:

```
pi@raspberrypi ~ $ ps -aux | less
```

Sometimes you may want to kill a rogue or unresponsive process. To do that, use `ps` to find its id, then use `kill` to stop it.

```
pi@raspberrypi ~ $ kill 95689
```

In the case of some system processes, you won't have permission to kill it (though you'll read about `sudo` in a moment).

Sudo and Permissions

Linux is a multiuser operating system; the general rule is that everyone owns their own files and can create, modify, and delete them within their own space on the filesystem. The root (or super) user can change any file in the filesystem, which is why it is good practice to not log in as root on a day-to-day basis.

As the pi user there's not much damage you can do to the system. As superuser you can wreak havoc, accidentally or by design. Be careful when using sudo, especially when moving or deleting files. Of course, if you things go badly, you can always make a new SD card image (see Appendix A).

There are some tools like sudo that allow users to act like super users for performing tasks like installing software without the dangers (and responsibilities) of being logged in as root. You'll be using sudo a lot when interacting with hardware directly, or when changing system-wide configurations.

Each file belongs to a particular user and a particular group. Use the chown and chgrp commands to change the owner or group of a file. Note that you must be root to use either of these two commands:

```
pi@raspberrypi ~ $ sudo chown pi garply.txt
pi@raspberrypi ~ $ sudo chgrp staff plugh.txt
```

Each file also has a set of *permissions* that show whether a file can be read, written or executed. These permissions can be set for the owner of the file, the group, or for everyone (see Figure 2-5).

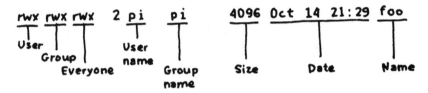

Figure 2-5. *File permissions for owner, group, and everyone.*

You set the individual permissions with the chmod command. The switches for chmod are summarized in Table 2-2.

Table 2-2. *The switches that can be used with chmod.*

u	user
g	group
o	others not in the group
a	all/everyone
r	read permission
w	write permission
x	execute permission
+	add permission
-	remove permission

Here are a few examples of how you can combine these switches :

```
chmod u+rwx,g-rwx,o-rwx wibble.txt ❶
chmod g+wx wobble.txt ❷
chmod -rw,+r wubble.txt ❸
```

❶   Allow only the user to read, write, execute

❷   Add permission to read and execute to entire group

❸   Make read only for everyone

The only thing protecting your user space and files from other people is your password, so you better chose a strong one. Use the `passwd` command to change it, especially if you're putting your Pi on a network.

The Network

Once you're on a network, there are a number of Linux utilities that you'll be using on a regular basis. When you're troubleshooting an Internet connection, use `ifconfig`, which displays all of your network interfaces and the IP addresses associated with them (see Figure 2-6).

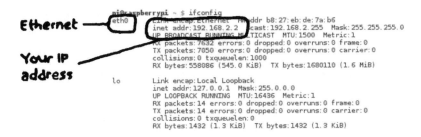

Figure 2-6. *The ifconfig command gives you information about all of your network interfaces.*

The `ping` command is actually the most basic tool for troubleshooting network conventions. You can use ping (think sonar) to test whether there is a two way connection between two IP addresses on the network or Internet. Note that many web sites block ping traffic, so you may need to ping multiple sites to accurately test a connection:

```
ping yahoo.com
ping altavista.com
ping lycos.com
ping netscape.com ❶
```

❶   Fail happens here.

To log in to another computer remotely (and securely, with encrypted pass-words), you can use the Secure Shell (ssh). The computer on the remote side needs to be running an SSH server for this to work, but the ssh client comes built into Raspbian. In fact, this is a great way to work on your Raspberry Pi without a monitor or keyboard, as discussed later in "Running Headless" (page 31).

Related to SSH is the `sftp` program, which allows you to securely transfer files from one computer to another. Rounding out the set is `scp`, which you can use to copy files from one computer to another over a network or the Internet. The key to all of these tools is that they use the Secure Sockets Layer (SSL) to transfer files with encrypted login information. These tools are all standard stalwart Linux tools.

Running Headless

If you want to work on the Raspberry Pi without plugging in a monitor, key-board, and mouse, there are some ways to set it up to run *headless*. If all you require is to get in to the command line, you can simply hook the Raspberry Pi up to the network and use an ssh client to connect to it (user name: pi, password: raspberry). The Terminal utility on the Mac will do, PuTTY (*http:// www.chiark.greenend.org.uk/~sgtatham/putty/*) on Windows or Linux, or `ssh` on Linux. The ssh server on the Raspberry Pi is enabled by default (run `raspi-config` again if for some reason it doesn't launch at startup).

Another way to connect to the Pi over a network connection is to run a Virtual Network Computing Server (VNC server) on the Pi and connect to it using a VNC client. The benefit of this is that you can run a complete working graph-ical desktop environment in a window on your laptop or desktop. This is a great solution for a portable development environment. See the Raspberry Pi Hub (*http://elinux.org/RPi_VNC_Server*) for extensive instructions on how to install TightVNC (*http://www.tightvnc.com/*), a lightweight VNC server.

/etc

The */etc* directory holds all of the system-wide configuration files and startup scripts. When you ran the configuration scripts the first time you started up, you were changing values in various files in the /etc directory. You'll need to invoke super user powers with sudo to edit files in */etc*; if you come across some tutorial that tells you to edit a configuration file, use a text editor to edit and launch it with sudo:

```
pi@raspberrypi ~ $ sudo nano /etc/hosts
```

Setting the Date and Time

A typical laptop or desktop will have additional hardware and a backup battery (usually a coin cell) to save the current time and date. The Raspberry Pi does not, but Raspbian is configured to automatically synchronize its time and date with an Network Time Protocol (NTP) server when plugged into a network.

Having the correct time can be important for some applications (see the example in Chapter 7 using cron to control a lamp). To set the time and date manually, use the date program:

```
$ sudo date --set="Sun Nov 18 1:55:16 EDT 2012"
```

Installing New Software

One of the areas that Linux completely trounces other operating systems is in software package management. *Package managers* handle the downloading and installation of software, and they automatically handle downloading and installing dependencies. Keeping with the modular approach, many software packages on Linux depend on other pieces of software. The package manager keeps it all straight, and the package managers on Linux are remarkably robust.

Raspbian comes with a pretty minimal set of software, so you will soon want to start downloading and installing new programs. The examples in this book will all use the command line for this, since it is the most flexible and quickest way of installing software.

The program apt-get with the -install switch is used to download software. apt-get will even download all of the other software that your package requires so you don't have to go hunting around for dependencies. Software has to be installed with superuser permissions, so always use sudo (this command installs the Emacs text editor):

```
pi@raspberrypi ~ $ sudo apt-get install emacs
```

 Taking a Screenshot

One of the first things we needed to figure out when writing this book was how to take screenshots on the Pi. We found a program called scrot (an abbreviation for SCReenshOT). Another option to capture screenshots is to install the GNU Image Manipulation Program (Gimp) or ImageMagick, but scrot worked for us. To install scrot type:

```
sudo apt-get install scrot
```

Going Further

There's much more to Linux, and many places to learn more. Some good starting points are:

Linux Pocket Guide (http://shop.oreilly.com/product/0636920023029.do)
Handy as a quick reference.

Linux in a Nutshell (http://shop.oreilly.com/product/9780596154493.do)
More detailed, but still a quick reference guide.

The Debian Wiki (http://wiki.debian.org/FrontPage)
Raspbian is based on Debian, so a lot of the info on the Debian wiki applies to Raspbian as well.

The Jargon File (http://catb.org/jargon/)
Also published as the *New Hacker's Dictionary*, this collection of definitions and stories is required reading on the Unix/Linux subculture.

3/Python On The Pi

Python is a great first programming language; it's clear and easy to get up and running. More important, there are a lot of other users to share code with and ask questions.

Guido van Rossum created Python, and very early on recognized its use as a first language for computing. In 1999, van Rossum put together a widely-read proposal called Computer Programming for Everybody (*http://www.python.org/doc/essays/cp4e.html*) that laid out a vision for an ambitious program to teach programming in the elementary and secondary grade schools using Python. More than a decade later, it looks like it is actually happening with the coming of the Raspberry Pi.

Python is an *interpreted language*, which means that you can write a program or script and execute it directly rather than compiling it into machine code. Interpreted languages are a bit quicker to program with, and you get a few side benefits. For example, in Python you don't have to explicitly tell the computer whether a variable is a number, a list, or a string; the interpreter figures out the data types when you execute the script.

The Python interpreter can be run in two ways: as an interactive shell to execute individual commands, or as a command line program to execute standalone scripts. The integrated development environment (IDE) bundled with Python and the Raspberry Pi is called IDLE (see Figure 3-1).

The Python Version Conundrum

The reason you see two versions of IDLE is that there are two versions of Python installed on the Pi. This is common practice (though a bit confusing). As of this writing, Python 3 is the newest, but changes made to the language between versions 2 and 3 made the latter not backward compatible. Even though Python 3 has been around for years it took a while for it to be widely adopted, and lots of user contributed packages have not been upgraded to Python 3. It gets even more confusing when you search the Python documentation; make sure you're looking at the right version!

The examples in this book will work with Python 2.7 or 3.X, unless otherwise noted.

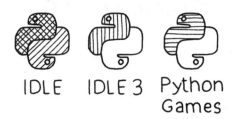

IDLE   IDLE 3   Python
Games

Figure 3-1. *Python options on the Raspbian Desktop: The IDLE integrated development environment for Python 2.0 (left), IDLE for Python 3.0 (middle), and a collection of sample games implemented in Pygame from invent-withpython.com (right).*

Hello Python

The best way to start learning Python is to jump right in. Although you can use any text editor to start scripting, we'll start out using the IDE. Open up the IDLE 3 application. To run IDLE, double-click the IDLE 3 icon on the desktop, or click the desktop menu in the lower left, and choose Programming→IDLE 3.

IDLE can take several seconds to start up, but when it appears, you'll see a window with the interactive shell. The triple chevron (>>>) is the interactive prompt; when you see the prompt, it means the interpreter is waiting for your commands. At the prompt type:

```
>>> print("Saluton Mondo!")
```

And hit Enter or Return. Python executes that statement and you'll see the result in the shell window. Note that the print() command is one of the things that changed in Python 3.0; if you get a syntax error, check to make sure you're running the 3.0 version of IDLE.

You can use the shell as a kind of calculator to test out statements or calculations:

```
>>> 3+4+5
12
```

Think of the statements executed in the interactive shell as a program that you're running one line at a time. You can set up variables or import modules:

```
>>> import math
>>> (1 + math.sqrt(5)) / 2
1.618033988749895
```

The import command makes all of Python's math functions available to your program (more about modules in "Objects and Modules" (page 40)). To set up a variable, use the assignment operator (=):

```
>>> import math
>>> radius = 200
>>> radius * 2 * math.pi
125.66370614359173
```

If you want to clear all variables and start in a fresh state, select Shell→Restart Shell to start over. You can also use the interactive shell to get information about how to use a particular statement, module, or other Python topics with the help() command:

```
help("print")
```

To get a listing of all of the topics available, try:

```
help("topics")
help("keywords")
help("modules")
```

The Python interpreter is good for testing statements or simple operations, but you will often want to run your Python script as you would a standalone application. To start a new Python program, select File→New Window, and IDLE will give you a script editing window (see Figure 3-2).

Figure 3-2. *The IDLE interactive shell (left) and an editor window (right).*

Try typing a line of code and selecting Run→Run Module. You'll get a warning that "Source Must Be Saved OK To Save?". Save your script in your home directory as *SalutonMondo.py* and you'll see it execute in the shell.

Sometimes you may not want to use the IDLE environment. To run a script from the command line, open up the LXTerminal and type:

```
python SalutonMondo.py
```

That's really all of the basic mechanics you need to know to get up and running with the environment. Next you'll need to start learning the language.

Command Line vs. IDLE

One thing you will notice is that the output of IDLE is very slow when running example code that prints to the shell. To get an idea how slow, keep IDLE open and open a new LXTerminal alongside. Save the script as *CountEvens.py* in IDLE and type the following at the command line prompt:

```
python CountEvens.py
```

Actually, before you do that, run the same script in IDLE using Run Module; it will need the head start. You'll quickly get an idea of the overhead from using the IDE on the fairly limited resources of the Pi. The examples later in the book will all be executed from the command line, but IDLE can still be used as an editor if you like.

A Bit More Python

If you're coming to Python from the Arduino world, you're used to writing programs (known as *sketches* in Arduino, but often called *scripts* in Python) in a setup/loop format, where **setup()** is a function run once and **loop()** is a function that executes over and over. The following example shows how to achieve this in Python. Select New Window from the shell in IDLE 3 and type the following:

```
# Setup
n = 0

# Loop
while True:
    n = n + 1
    # The % is the modulo operator
    if ((n % 2) == 0):
        print(n)
```

Select Run Module and give your script a name (such as *EvenIntegers.py*). As it runs, you should see all even integers printed (press **Control-C** to interrupt the output, because it will go on forever).

In the example above, you may not notice that each level of indentation is four spaces, not a tab (but you can press tab in IDLE and it will dutifully insert spaces for you). Indentation has structural meaning in Python. This is one of the big stumbling blocks for beginners, or when copying and pasting code. Still, we feel that the mandatory use of whitespace makes Python a fairly readable language. See the Python Style Guidelines (*http://www.python.org/dev/peps/pep-0008/#indentation*) for tips on writing readable code.

It is important to watch your whitespace; Python is a highly structured language where the whitespace determines the structure. In the next example, everything indented one level below the **loop()** function is considered part of that function. The end of a loop is determined by where the indentation moves up a level (or end of the file). This differs from languages like C that delimit blocks of code with brackets or other markers.

Use functions to put chunks of code into a code block that can be called from other places in your script. To rewrite the previous example with functions, do the following (when you go to run it, save it as *CountEvens.py*):

```
# Declare global variables
n = 0 ❶

# Setup function
def setup(): ❷
    global n
    n = 100

def loop(): ❸
    global n
    n = n + 1
    if ((n % 2) == 0):
        print(n)

# Main ❹
setup()
while True:
    loop()
```

In this example, the output will be every even number from 102 on. Here's how it works:

❶   First the variable **n** is defined as a global variable that can be used in any block in the script.

❷   Here, the **setup()** function is defined (but not yet executed).

❸   Similarly, here's the definition of the **loop()** function.

❹   In the main code block, **setup()** is called once, then **loop()**.

The use of the `global` keyword in the first line of each function is important; it tells the interpreter to use the global variable n rather than create a second (local, or private to that function) n variable usable only in the function.

This tutorial is too short to be a complete Python reference. To really learn the language, you may want to start with Think Python (*http://shop.oreil ly.com/product/0636920025696.do*) and the Python Pocket Reference (*http://shop.oreilly.com/product/9780596158095.do*). The rest of this chapter will give you enough context to get up and running with the later examples and will map out the basic features and modules available. Chapter 4 covers the PyGame framework, a good way to create multimedia programs on the Pi.

Objects and Modules

You'll need to understand the basic syntax of dealing with objects and modules to get through the examples in this book. Python is a clean language, with just 34 reserved *keywords* (see Table 3-1). These keywords are the core part of the language that let you structure and control the flow of activity in your script. Pretty much everything that isn't a keyword can be considered an *object*. An object is a combination of data and behaviors that has a name. You can change an object's data, retrieve information from it, and even manipulate other objects.

Table 3-1. *Python has just 34 reserved keywords*

Conditionals	Loops	Built-in Functions	Classes, Modules, Functions	Error Handling
if	for	print	class	try
else	in	pass	def	except
elif	while	del	global	finally
not	break		lambda	raise
or	as		nonlocal	assert
and	continue		yield	with
is			import	
True			return	
False			from	
None				

In Python, strings, lists, functions, modules, and even numbers are objects. A Python object can be thought of as an encapsulated collection of attributes and methods. You get access to these attributes and methods using a simple dot syntax. For example, type this at the interactive shell prompt to set up a string object and call the method that tells it to capitalize itself:

```
>>> myString = "quux"
>>> myString.capitalize()
'Quux'
```

Or use **reverse()** to rearrange a list in reverse order:

```
>>> myList = ['a', 'man', 'a', 'plan', 'a', 'canal']
>>> myList.reverse()
>>> print(myList)
['canal', 'a', 'plan', 'a', 'man', 'a']
```

Both String and List are built-in modules of the *standard library*, which are available from any Python program. In each case, the String and List modules have defined a bunch of functions for dealing with strings and lists, including **capital ize()** and **reverse()**.

Some of the standard library modules are not built-in and you need to explicitly say you're going to use them with the **import** command. To use the **time** module from the standard library to gain access to helpful functions for dealing with timing and timestamps, use:

```
import time
```

You may also see the use of **import as** to rename the module in your program:

```
import time as myTime
```

You'll also see the use of **from import** to load select functions from a module:

```
from time import clock
```

Here's a short example of a Python script using the **time** and **datetime** modules from the standard library to print the current time once every second:

```
from datetime import datetime
from time import sleep

while True:
    now = str(datetime.now())
    print(now)
    sleep(1)
```

The **sleep** function stops the execution of the program for one second. One thing you will notice after running this code is that the time will drift a bit each time. That's for two reasons:

1. The code doesn't take into account the amount of time it takes to calculate the current time (.9 seconds would be a better choice)

2. Other processes are sharing the CPU and may take cycles away from your program's execution. This is an important thing to remember: when programming on the Raspberry Pi you are not executing in a *real time environment*.

If you're using the sleep() function, you'll find that it is accurate to more then 5ms on the Pi.

Next, let's modify the example to open a text file and periodically log some data to it. Everything is a string when handling text files. Use the str() function to convert numbers to strings (and int() to change back to an integer).

```
from datetime import datetime
from time import sleep
import random

log = open("log.txt", "w")

while i in range(5):
    now = str(datetime.now())
    # Generate some random data in the range 0-1024
    data = random.randint(0, 1024)
    log.write(now + " " + str(data) + "\n")
    print(".")
    sleep(.9)
log.flush()
log.close()
```

 In a real data logging application you'll want to make sure you've got the correct date and time set up on your Raspberry Pi, as described in Chapter 1.

Here's another example (*ReadFile.py*) that reads in a filename as an argument from the command line (run it from the shell with python3 Read File.py *filename*. The program opens the file, reads each line as a string and prints it. Note that print() acts like println() does in other languages; it adds a newline to the string that is printed. The end argument to print() suppresses the newline.

```
# Open and read a file from command line argument
import sys

if (len(sys.argv) != 2):
    print("Usage: python ReadFile.py filename")
    sys.exit()

scriptname = sys.argv[0]
filename = sys.argv[1]

file = open(filename, "r")
```

```
lines = file.readlines()
file.close()

for line in lines:
    print(line)
```

Even More Modules

One of the reasons Python is so popular is that there are a great number of user-contributed modules that build on the standard library. The Python Package Index (PyPI) (*http://pypi.python.org/pypi*) is the definitive list of packages (or modules); some of the more popular modules that could be particularly useful on the Raspberry Pi are shown in Table 3-2. You'll be using some of these modules later on, especially the GPIO module to access the general inputs and outputs of the Raspberry Pi.

Table 3-2. *Some packages of particular interest to Pi users*

Module	Description	URL	Package Name
RPi.GPIO	Access to GPIO pins	*http:// code.google.com/p/ raspberry-gpio-python/*	python-rpi.gpio
Pygame	Gaming framework	*http://pygame.org*	python-pygame
SimpleCV	Easy API for Computer Vision	*http://simplecv.org/*	No package
Scipy	Scientific computing	*http://www.scipy.org/*	python-scipy
Numpy	The numerical underpinings of Scipy	*http://numpy.scipy.org/*	python-numpy
Flask	Microframework for web development	*http://flask.pocoo.org/*	python-flask
Requests	"HTTP for Humans"	*http://ocs.python-requests.org*	python-requests
PIL	Image processing	*http://www.python ware.com/products/pil/*	python-imaging
wxPython	GUI framework	*http://wxpython.org*	python-wxgtk2.8
PySerial	Access to serial port	*http://pyserial.source forge.net/*	python-serial
pyUSB	FTDI-USB interface	*http://bleyer.org/pyusb*	No package

To use one of these you'll need to download the code, configure the package, and install it. The `serial` module can be installed this way, for example:

```
sudo apt-get install python-serial
```

If a package has been bundled by its creator using the standard approach to bundling modules (with Python's distutils tool), all you need to do is download the package, uncompress it and type:

```
python setup.py install
```

You may also want to look at the Pip package installer (*http://www.pip-installer.org*), a tool that makes it quite easy to install packages from the PyPI.

Troubleshooting Errors

Inevitably you'll run into trouble with your code and you'll need to track down and squash a bug. The IDLE interactive mode can be your friend; the Debug menu provides several tools that will help you understand how your code is actually executing. You also have the option of seeing all your variables and stepping through the execution line by line.

Syntax errors are the easiest to deal with; usually this is just a typo or a misunderstood aspect of the language. *Semantic errors* — where the program is well formed but doesn't perform as expected — can be harder to figure out. That's where the debugger can really help unwind a tricky bug. Effective debugging takes years to learn, but here is a quick checklist of things to check when programming the Pi in Python:

- Use `print()` to show when the program gets to a particular point.
- Use _print()+ to show the values of variables as the program executes.
- Double check whitespace to make sure blocks are defined the way you think they are.
- When debugging syntax errors, remember that the actual error may have been introduced well before the interpreter reports it.
- Double check all of your global and local variables
- Check for matching parentheses
- Make sure the order of operations is correct in calculations; insert parentheses if you're not sure. For example, `3 + 4 * 2` and `(3 + 4) * 2` yield different results.

After you're comfortable and experienced with Python, you may want to look at the `code` and `logging` modules for more debugging tools.

Going Further

There is a lot more to Python, and here are some resources that you'll find useful:

*Think Python (http://shop.oreilly.com/product/0636920025696.do) by Al-
len Downey*

This is a clear and fairly concise approach to programming (that happens to use Python).

*The Python Pocket Reference (http://shop.oreilly.com/product/
9780596158095.do)*

Because sometimes flipping through a book is better than clicking
through a dozen Stack Overflow posts.

Stack Overflow (http://stackoverflow.com/)

That said, Stack Overflow is an excellent source of collective knowledge.
It works particularly well if you're searching for a specific solution or error
message; chances are someone else has had the same problem and
posted here.

Learn Python the Hard Way (http://learnpythonthehardway.org/)

A great book and online resource; at the very least read the introduction
"The Hard Way Is Easier."

*Python For Kids (http://shop.oreilly.com/product/9781593274078.do) by
Jason R. Briggs*

Again, more of a general programming book that happens to use Python
(and written for younger readers).

4/Animation and Multimedia in Python

Pygame is a lightweight framework for creating simple games in Python. You can also think of it as a tool for general multimedia programming; it's a convenient way to just draw graphics on the screen, play sounds, or handle keyboard and mouse events.

Pygame is a software wrapper around another library called the Simple DirectMedia Layer (SDL). SDL handles all of the low-level access to the Pi's keyboard, mouse, audio drivers and video drivers. Pygame simplifies SDL even further.

This focus of this chapter is more on the basic multimedia capabilities of Pygame, rather than a tutorial on game programming. Other resources for game programming are provided at the end.

Hello Pygame

Pygame comes pre-installed on your Raspberry Pi. There's a version of Pygame that works with Python 3.0, but the version that comes with Raspbian only works with version 2.7. You can either switch over to the non-3.0 version of IDLE (just double-click the IDLE icon instead of the IDLE 3 icon), or just run your script from the command line and the 2.7 interpreter will be used.

The following example shows the minimal number of steps to create a Pygame program; it draws a red circle in a new window:

```
import pygame ❶

width = 640 ❷
height = 480
radius = 100
fill = 1

pygame.init() ❸

window = pygame.display.set_mode((width, height)) ❹
```

```
window.fill(pygame.Color(255, 255, 255)) ❺

while True: ❻
    pygame.draw.circle(window, ❼
                       pygame.Color(255, 0, 0),
                       (width/2, height/2),
                       radius, fill)
    pygame.display.update() ❽
```

❶ Import the Pygame module, which makes all of the Pygame objects and functions available for use.

❷ Set up some global variables here.

❸ Always call the `init()` function first; this performs some steps needed to initialize Pygame before you use it, and will also call the `init()` function for all of the Pygame submodules.

❹ Set up the window, which is a Pygame *Surface* object, an area on which to draw.

❺ Fill the window with white.

❻ Loop forever. You can think of each iteration of this loop as a single frame of an infinite animation.

❼ Draw a filled circle in red at the center of the window.

❽ All of the drawing commands in the loop make up one frame of the animation and are drawn into an offscreen buffer that is hidden. When you're done drawing the frame, call `display.update()` to update the image on the screen.

In all of these examples, you'll have to use Control-C on the command line to kill the script when you're done. If you want to enable the close button on the Pygame window, add the following code to the end of the while loop:

```
while True:
    if pygame.QUIT in [e.type for e in pygame.event.get()]:
        break
```

That will catch the event triggered when the button is pressed. For more on events see "Handling Events and Inputs" (page 51).

Pygame consists of a collection of submodules and objects; the basics of how they are used are described in the rest of this chapter.

Pygame Surfaces

A Pygame Surface is just a rectangular image; Surfaces are combined and layered to create each scene in a frame of the game or animation. The pixels of a Surface are represented by a sequence of three 8-bit RGB numbers; e.g. (0, 255, 0) represents green. Add a fourth number for transparency: (0, 255, 0, 127) is 50% transparent.

The display window is the base Surface on which all other Surfaces are drawn. The `pygame.display` module controls and provides information about the display window. Use the `set_mode()` function to create a new display window, and the `update()` function redraw the display every frame.

To load an image from a file into a Surface to be displayed, use the `load()` function from the `pygame.image` module. Once you have a display Surface you can add images to it by creating a new Surface object and using the `blit()` function to combine them:

```
import pygame

pygame.init()
screen = pygame.display.set_mode((450, 450))
background = pygame.image.load("background.png") ❶
background.convert_alpha() ❷
screen.blit(background, (0, 0)) ❸
while True:
    pygame.display.update()
```

❶ Loads a background image that needs to be in the same directory as the Python program.

❷ The `convert_alpha()` function changes the format of the Surface to match the current display. This isn't required, but is suggested as it will speed image compositing.

❸ By default the screen Surface will be black. The `blit()` function combines the background Surface with a black background.

Here's an example of combining two images (see Figure 4-1):

```
import pygame

pygame.init()
screen = pygame.display.set_mode((450, 450))
background = pygame.image.load("background.png").convert_alpha()
theremin = pygame.image.load("theremin.png").convert_alpha()
screen.blit(background, (0, 0))
screen.blit(theremin, (135, 50))
while True:
    pygame.display.update()
```

Figure 4-1. *Blitting two images together.*

The `pygame.transform` module provides functions for rotating and scaling surfaces. You can get access to the individual pixels of a Surface using the functions of the `pygame.surfarray` module.

Surfaces are always rectangular images when created; even though the transparent areas of the previous example Surface were combined to make it look non-rectangular, the Surface is still a rectangle. If you want a Surface with a non-rectangular boundary (for pixel-perfect collision detection, for example) you can set a mask from another surface with the `pygame.mask` module. Masks are described in more detail in the game programming resources at the end of the chapter.

Drawing on Surfaces

You saw an example of the pygame circle drawing function previously; the `pygame.draw` module provides several more for drawing rectangles, lines, arcs, and ellipses. The `pygame.gfxdraw` module is another way of drawing shapes. It provides a few more options but is considered experimental and the API may change.

To draw text on the screen you first create a new *Font* object (provided by the `pygame.font` module), then use that to load a font and render the text. To find a list of all of the fonts available on your Raspberry Pi, use the `pygame.font.get_fonts()` function:

```
import pygame

pygame.init()
for fontname in pygame.font.get_fonts():
    print fontname
```

As you can see, there are few fonts available with the base Raspbian distribution. Use the SysFont object to render some text using the *freeserif* font:

```
import pygame

pygame.init()
screen = pygame.display.set_mode((725, 92))
font = pygame.font.SysFont("freeserif", 72, bold = 1)
textSurface = font.render("1 Theremin Per Child!", 1,
                          pygame.Color(255, 255, 255))
screen.blit(textSurface, (10, 10))
while True:
    pygame.display.update()
```

 If you want more fonts, try:

```
sudo apt-get install ttf-mscorefonts-installer
sudo apt-get install ttf-liberation
```

Handling Events and Inputs

In Pygame, user-initiated events like pressing a key or clicking or moving the mouse are all captured as an *Event* object and put in the *event queue*, where they accumulate until your program does something with them. The py game.event module provides functions for handling all the events that have piled up since the last time you looked at the queue. You can even create your own event types to implement a messaging system. Below is a simple example that uses the event queue to expand on the red circle program; in each frame, the circle is redrawn at the current mouse location and the radius gets larger the closer it is to the edge of the window (see Figure 4-2).

Figure 4-2. *Output of the Pygame events example.*

```
import pygame
from pygame.locals import * ❶

width, height = 640, 640
radius = 0
mouseX, mouseY = 0, 0 ❷

pygame.init()
window = pygame.display.set_mode((width, height))
window.fill(pygame.Color(255, 255, 255))

fps = pygame.time.Clock() ❸

while True: ❹
    for event in pygame.event.get(): ❺
        if event.type == MOUSEMOTION: ❻
            mouseX, mouseY = event.pos
        if event.type == MOUSEBUTTONDOWN: ❼
            window.fill(pygame.Color(255, 255, 255))
    radius = (abs(width/2 - mouseX)+abs(height/2 - mouseY))/2 + 1 ❽
    pygame.draw.circle(window, ❾
                        pygame.Color(255, 0, 0),
                        (mouseX, mouseY),
                        radius, 1)
    pygame.display.update()
    fps.tick(30) ❿
```

❶ The `pygame.locals` module defines a number of constants like `MOUSE MOTION`. Importing it here allows us to use these constants without prepending `pygame.` to them.

❷ Variables to store the mouse coordinates.

❸ This function initializes an object that we'll use as a frame counter. With this `fps` variable (Frames Per Second) you can wait a certain amount in each frame to achieve a regular framerate.

❹ Loop forever; each time through the loop is one frame.

❺ Loop through the event queue. Each time through, the event variable will hold the next event in the queue.

❻ If the event is a mouse movement, update the variable holding the mouse location.

❼ If the event is a mouse click, clear the screen.

❽ Vary the radius based on how far away from the center the mouse is.

❾ Draw the circle.

❿ Wait so that the frame rate is 30 fps.

Other modules that are handy for dealing with events and user input are:

pygame.time
> A module for monitoring time.

pygame.mouse
> A module to get information about the mouse.

pygame.key
> A module to get information about the keyboard, and many constants representing the keys.

pygame.joystick
> A module for working with a joystick.

Going Full Screen

To run your program in a window that takes over the whole screen, set the `pygame.FULLSCREEN` flag when you set the display mode. When in full screen mode, make sure to have some way of getting out of the script, since you won't be able to kill it with `Control-C`:

```python
import pygame
import random
from time import sleep

running = True
pygame.init()
screen=pygame.display.set_mode((0,0), pygame.FULLSCREEN)
while running:
    pygame.draw.circle(
        screen,
        pygame.Color(int(random.random()*255),
                     int(random.random()*255),
                     int(random.random()*255)),
        (int(random.random()*1500),
         int(random.random()*1500)),
        int(random.random()*500), 0)
    pygame.display.update()
    sleep(.1)
    for event in pygame.event.get():
        if event.type == pygame.KEYDOWN:
            running = False
pygame.quit()
```

The script will exit when any key on the keyboard is pressed.

Sprites

Most of the movable and controllable graphical elements of a game will be handled as *sprites*. The `pygame.sprite` module provides all of the basic functions to draw sprites on the screen and handle collisions between them. Sprites can also be grouped to be controlled and updated together. A complete working game using sprites is beyond the scope of this book (the "Further Reading" (page 58) section has some more resources).

Sprites are best used when you'll be creating several screen elements that share a lot of the same code. Here is an example of how to handle creating and updating a couple of sprites; the result is two balls that bounce off the sides of the screen. You can add other balls by creating a new sprite with a different starting coordinate, direction, and speed.

```
import pygame

class Ball(pygame.sprite.Sprite): ❶

    def __init__(self, x, y, xdir, ydir, speed): ❷
        pygame.sprite.Sprite.__init__(self)
        self.image = pygame.Surface([20, 20])
        self.image.fill(pygame.Color(255, 255, 255))
        pygame.draw.circle(self.image,
                            pygame.Color(255,0,0),
                            (10,10), 10, 0)
        self.rect = self.image.get_rect()
        self.x, self.y = x, y ❸
        self.xdir, self.ydir = xdir, ydir
        self.speed = speed

    def update(self): ❹
        self.x = self.x + (self.xdir * self.speed)
        self.y = self.y + (self.ydir * self.speed)
        if (self.x < 10) | (self.x > 490):
            self.xdir = self.xdir * -1
        if (self.y < 10) | (self.y > 490):
            self.ydir = self.ydir * -1
        self.rect.center = (self.x, self.y)

pygame.init()
fps = pygame.time.Clock()
window = pygame.display.set_mode((500, 500))
ball = Ball(100, 250, 1, 1, 5) ❺
ball2 = Ball(400, 10, -1, -1, 8)

while True:
    ball.update() ❻
    ball2.update()
    window.fill(pygame.Color(255,255, 255))
```

```
window.blit(ball.image, ball.rect) ❼
window.blit(ball2.image, ball2.rect)
pygame.display.update()
fps.tick(30)
```

❶ The *class* statement creates a new object (a Ball) that is based on the Sprite object. You can then define your own functions for drawing the sprite, and how the sprite behaves each time it is updated.

❷ This function is called when the `Ball()` function is called to create a new ball.

❸ These variables are stored as part of each *instance* of Ball. Each Ball in essence carries around its own set of variables.

❹ The `update()` function is called with each frame. The saved position is moved based on the saved direction and speed, then tested to see if it has come near the edge. If so, the direction on that particular axis is reversed.

❺ Create two balls with different starting locations, directions, and speed.

❻ Update the ball with a new location based on its saved position, direction, and speed.

❼ Draw the ball at its current location.

Playing Sound

In Pygame you can load sound files and play them back using the `pygame.mix er` module, or you can use the `pygame.midi` module to send MIDI events to other software running on the Pi or MIDI hardware hooked up to the USB port. The following example plays back a WAV-formatted sound sample (which you can download from the Internet Archive (*http://archive.org/ details/WilhelmScreamSample*)).

```
import pygame.mixer
from time import sleep

pygame.mixer.init(48000, -16, 1, 1024)

sound = pygame.mixer.Sound("WilhelmScream.wav")
channelA = pygame.mixer.Channel(1)
channelA.play(sound)
sleep(2.0)
```

An audio file (in the WAV format) is loaded and associated with a channel. The mixer actually plays back each channel in a separate process, so multiple sounds will play at the same time. The `sleep()` at the end is needed so the main process doesn't end before the sound has finished playing.

The mixer approach for playing back sound files is discussed further in Chapter 8 in the Simple Soundboard example. The MIDI capabilities of Pygame have intriguing possibilities; for example you can easily create a custom MIDI controller. As of this writing, the state of software synthesizers that work on Raspbian is kind of rough, and the analog audio output is not as good as the HDMI audio output. That will improve, however, and you can always use an external USB MIDI device. To find information about the MIDI devices connected to your Raspbery Pi, use:

```
import pygame
import pygame.midi  ❶

pygame.init()
pygame.midi.init()  ❷
for id in range(pygame.midi.get_count()):  ❸
    print pygame.midi.get_device_info(id)  ❹
```

❶   The midi module is not imported with the base pygame module.

❷   It also must be initialized separately.

❸   The get_count() function returns the number of MIDI-capable devices connected to the Pi either by USB, software synthesizers, or other virtual MIDI devices.

❹   Prints information about the device.

If you plug in an external MIDI keyboard you'll get something that looks like this for output:

```
('ALSA', 'Midi Through Port-0', 0, 1, 0)
('ALSA', 'Midi Through Port-0', 1, 0, 0)
('ALSA', 'USB Uno MIDI Interface MIDI 1', 0, 1, 0)
('ALSA', 'USB Uno MIDI Interface MIDI 1', 1, 0, 0)
```

The first string tells you that you're using the Advanced Linux Sound Architecture (ALSA), and the second tells describes each MIDI port. The last three numbers indicate whether the port is an input device, output device, and whether it is open or not. You can think of the example above as four different MIDI ports labeled 0 through 3:

Midi Through Port-0 (0, 1, 0)
    Port 0, an output used to talk to any software synthesizers you may have running on the Pi.

Midi Through Port-0 (1, 0, 0)
    Port 1, an input used to take MIDI controls from any software controllers you may have running on the Pi.

USB Uno MIDI Interface MIDI 1 (0, 1, 0)
Port 2, an output to an external USB MIDI interface hooked up to a keyboard.

USB Uno MIDI Interface MIDI 1 (1, 0, 0)
Port 3, an input from an external USB MIDI interface hooked up to a keyboard.

With this is mind, you can hook up an external keyboard with a USB interface and use Pygame to control it, as in the following example:

```
import pygame
import pygame.midi
from time import import sleep

instrument = 0 ❶
note = 74
volume = 127

pygame.init()
pygame.midi.init()

port = 2  ❷
midiOutput = pygame.midi.Output(port, 0)
midiOutput.set_instrument(instrument)
for note in range(0, 127):
    midiOutput.note_on(note, volume) ❸
    sleep(.25)
    midiOutput.note_off(note, volume)
del midiOutput
pygame.midi.quit()
```

❶   These are MIDI values, typically in the range 0 to 127

❷   Open port 2, the ouput associated with the USB MIDI controlled keyboard.

❸   Send a note on control signal, wait, then turn the note off.

It's clear that the Raspberry Pi has a lot of potential as a platform for music making.

Playing Video

Pygame can also be used to play videos using the the `pygame.movie` module. Video files must be encoded in the MPEG1 format. If you have a video in a different format, try the *ffmpeg* utility to convert between formats (you'll have to run `sudo apt-get install ffmpeg` first). To play it back, simply create a new Movie object and call the `play()` function:

```
import pygame
from time import sleep
pygame.init()
screen = pygame.display.set_mode((320,240))
movie = pygame.movie.Movie("foo.mpg")
movie.play()
while True:
    if not(movie.get_busy()):
        print("rewind")
        movie.rewind()
        movie.play()
    if pygame.QUIT in [e.type for e in pygame.event.get()]:
        break
```

If the video has an audio track you'll need to close Pygame's audio mixer before playing the movie. To tell the mixer to quit include this line before you play:

```
pygame.mixer.quit()
```

 Even More Examples

There's a whole Pygame module dedicated to more complete example programs: `pygame.examples`. You can find the source code for these examples in the */usr/share/pyshared/pygame/ examples* directory.

Further Reading

Pygame official documentation (http://www.pygame.org/docs/)
The official documentation is a bit sparse in places, but hopefully you'll be able to navigate it after reading this chapter.

Making Games with Python & Pygame (http://inventwithpython.com/) and Invent Your Own Computer Games with Python (http://inventwithpy thon.com/) by Al Sweigart
Two Creative Commons-licensed books; the games developed in these books are bundled with the Raspberry Pi.

5/Scratch on the Pi

Scratch was developed by the MIT Media Lab's Lifelong Kindergarten group as a new way of teaching programming to young people. Programs are constructed from colorful blocks, each of which performs an operation. The self-contained blocks eliminate the syntax problems that stymie many first timers using text-based programming languages.

To say that Scratch is not a powerful programming language misses the point, which is that it is a friendly environment for creating and making things happen quickly. A young programmer can see the blocks of code highlight as they're executed, and blocks can be changed and the effects seen in real time.

As you'll see, all Scratch programs are aimed at manipulating sprites on a *stage*. There's a large community of Scratch users, and the ability to share sprites and code with the community is baked right into the platform.

Hello Scratch

To show how easy it is to get started programming in Scratch, we'll start right in with a very simple "cat in the box" program. When you open up Scratch you'll see a single window open with several panes. The role of each pane is labelled in Figure 5-1. You can start Scratch by double-clicking its desktop icon, or by clicking the desktop menu in the lower left and choosing Education→Scratch.

Every Scratch program involves *Sprites* interacting on the *Stage*. Sprites are controlled by *scripts* that are built with a collection of blocks that are pulled from the *Block Palette*. When you open a new document you'll get the default Scratch cat sprite. If you don't want the cat, you can delete it (right click on the sprite in the *Sprite List* for the menu), then draw your own or grab a random surprise sprite created by someone else in the Scratch community.

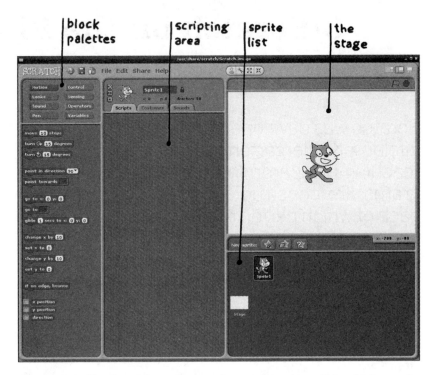

Figure 5-1. *The Scratch environment; everything is contained in one window, with subpanels for the block palettes, the scripting and costume area, the sprite list and the stage.*

The *Scripting Area* shows the script belonging to the currently selected sprite. Click on the cat in the sprite list and you'll see an empty scripting area for the sprite. You can build a script anywhere in this frame. If you create another sprite, you'll need to click on it to display its script.

Scripts are built up by dragging blocks from the Palette to the Scripting Area. Many blocks can contain other blocks, and how they fit together are indicated by their shape. There are three types of blocks in Scratch, as shown in Figure 5-2:

Hat blocks
> The *when green flag clicked* block is an example of a hat block, which sits at the top of a stack of blocks and waits for an event to happen.

Stackable blocks
> Blocks with an indentation on top and/or bump on the bottom fit together with other stackable blocks. The blocks are executed in top down order.

Reporter blocks

Reporter blocks have rounded or pointed edges, and fit inside the input areas of other blocks. Reporter blocks may be variables, or may provide information like the mouse coordinates or some condition.

Figure 5-2. *There are three types of blocks in Scratch: hat blocks (left), stackable blocks (center), and reporter blocks (right).*

Start a new script by selecting the *Control* block palette, and drag this block to the scripting area:

The next step is to use the stackable *forever* block that will execute all of the commands contained within it over and over under the script is stopped:

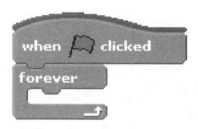

Be sure to practice reorganizing the blocks. When you pick up a block you also pick up all of the blocks attached after it. To separate a block you will often have to separate a whole stack of blocks, then grab the children of the block you want to remove. You can also right click on a block to duplicate it (and its children) and get help.

Next, select the Motion palette, and drag a *turn* block and place it in the forever loop.

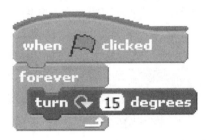

At this point you can click the green flag in the upper right of the stage and start executing the script. One of the neat things about Scratch is that you can make changes on the fly and see them immediately take effect on stage. This is a really good way to debug your program on the fly.

 When you click the green flag, a message is sent to all scripts in the project to start running (the stop sign sends a stop signal).

If you look at the *turn* block you'll see that the number of degrees to rotate is in a rounded rectangle shaped block. You can edit the angle directly, or you can replace it with another block of the same shape. You can have the sprite perform a random walk by making some changes.

First, replace the default value with the *pick random* block from the Operators palette. After you place the block, change the values so they select a random number between -10 and 10.

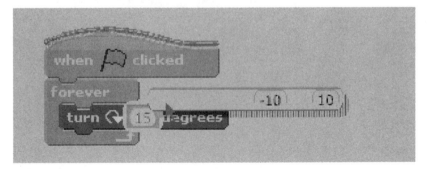

Now add a *move* block from the Motion palette, which will move the sprite a number of pixels in the direction it is currently facing. The direction is shown as a number and a blue line in the info bar above the scripting area. As soon as you place the block the sprite will start moving in a random walk.

Finally, to keep the sprite on the stage, add a *if on edge, bounce* block from the motion palette:

That's all there is to writing a Scratch script!

The Stage

The stage is the frame in the upper right corner where all of your sprites carry out their actions and interactions. Like sprites, the stage can have scripts that change its appearance or behavior. You can also paint backgrounds in the *Backgrounds* tab of the scripting area.

The coordinate system in Scratch is a little different than other multimedia environments like Pygame; it follows more of a math class model, where the origin (0, 0) is actually in the center of the stage. The visible part of the stage extends from (-240, 180) to (240, -180) as shown in Figure 5-3. Sprites can continue off stage if you let them, however; if you select a sprite you'll see the current position at the top of the script area, and the location of the cursor at the bottom of the stage.

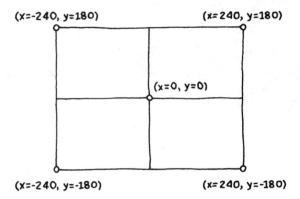

Figure 5-3. *The Scratch coordinate system, with the origin in the center of the stage.*

Two More Things to Know About Sprites

There are two additional tabs in the scripting area that you'll be using: *Costumes* and *Sounds*.

Costumes are a collection of images that the sprite can change into. Costumes can be used to create looping animations or to show different states that the sprite is in (for example, an exploding spaceship). For example, create a new sprite that is an open eye; we'll use changing costumes to make it blink. Select *Paint New Sprite* (it's the icon to the right of "New sprite" in the State area) and draw something like this, then click OK:

Select the *Costumes* tab, where you'll see the image you just drew. Change the name of the costume to *open*. Click the *Copy* button to create a new costume. Click *Edit* to erase the pupil and draw a closed eyelid, then click OK and change the name of this costume to *shut*:

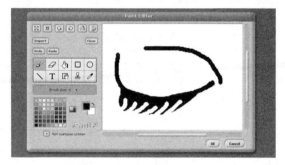

Next select the *Scripts* tab and drag these blocks to create a script that will blink the eye every second:

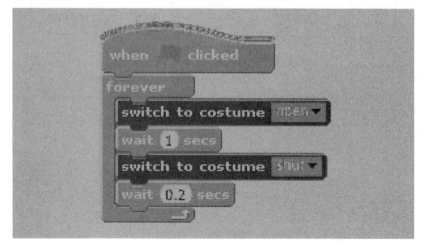

Finally, you can add sound effects to a sprite in the *Sounds* tab. Just hit record and you'll get the built-in Sound Recorder tool. You'll need an external USB sound card or microphone to be able to record sound. Give the sound effect a name and you can play it back using the *play sound* or *play sound until done* blocks.

A Bigger Example: Astral Trespassers

In the game Astral Trespassers, the player must shoot at incoming alien spaceships; the game is over when an alien hits the player's planet defending cannon. This simple game will show how all the pieces fit together.

First select File→New, then save the new game by selecting File→Save As. Next you'll need to create some sprites. Control-click or right click on the cat in the Sprite list and select Delete from the menu. Then create new sprites by painting your own or loading from an image file. I drew the 5 sprites in Figure 5-4 and scanned them in as a PNG.

You'll see that the sprites are added to the stage as you create them, and they are probably not the size you want them to be. Select the *Grow Sprite* or *Shrink Sprite* tool just above the Stage and click on the sprite to make it larger or smaller. You'll also want to rename the sprites to use the names shown in Figure 5-4.

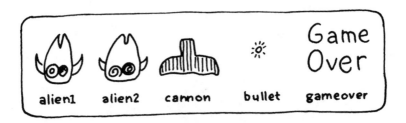

Figure 5-4. *The 5 sprites for Astral Trespassers.*

The aliens and the cannon each have two states; normal and exploded. We'll handle this by creating a costume for each state, as shown in Figure 5-5.

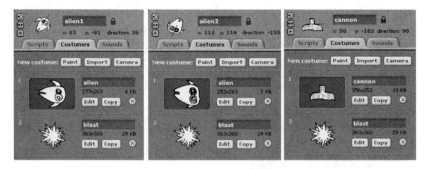

Figure 5-5. *Create two costumes for each sprite, one for the normal state and one for an explosion. Name the costumes as shown here.*

Each sprite will have its own script that will determine how it acts in the game. For clarity each action will be a complete standalone scripts that will all run at the same time.

 Appendix B shows the integrated scripts for each sprite, and also shows how to add support for multiple repeating bullets.

Let's start with the first alien. Select the sprite and drag the following blocks to the scripting area:

This block will move the sprite to its starting location and make sure it is visible and in its non-exploded state.

Next, drag and assemble the following blocks somewhere in the scripting area:

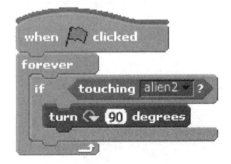

Hit the green flag and you'll see this is pretty much the cat in the box example from the beginning of the chapter. Since there are three sprites here, you'll need to handle what happens when the sprites collide. If two aliens collide, one way to make sure they don't overlap is to turn 90 degrees on collision:

If the bullet is touching the alien, we want it to explode:

```
when [flag] clicked
forever
    if  < touching bullet ? >
        switch to costume blast
        repeat 10
            change whirl effect by 25
        clear graphic effects
        hide
        switch to costume alien
        wait 3 secs
        go to x: (pick random 1 to 250) y: 185
        show
```

The explosion is handled by changing to the explosion costume, applying a special effect for a short time, then hiding the sprite which takes it off the stage. After a few seconds, the alien will return to a random location at the top of the screen.

That's all the scripting we need for the alien. If you have multiple sprites that do the same (or similar) thing, you can copy code between sprites by right clicking on a block and selecting *duplicate*. Drag the duplicated blocks to the second alien and repeat for each snippet of blocks. Now select the second alien and change the starting location and the collision test:

```
when [flag] clicked
switch to costume alien
go to x: 160 y: 185
show
```

```
when [flag] clicked
forever
    if  < touching alien1 ? >
        turn ↻ 90 degrees
```

Next select the cannon sprite. First move it into position and make sure it is visible and in the non-exploded state:

```
when 🏳 clicked
switch to costume cannon ▾
go to x: 0 y: -165
show
```

The cannon can only move left and right, controlled by the keyboard arrow keys). Here is the code to respond to keyboard presses:

```
when 🏳 clicked
forever
    if  key left arrow ▾ pressed?
        change x by -10

    if  key right arrow ▾ pressed?
        change x by 10
```

Test to see if the cannon is touching either alien with the *touching* block from the sensing palette. If it is, make the cannon explode in a similar manner to the aliens. After the explosion, the cannon will broadcast a message to all of the other sprites that the game is over. This message will be picked up by the Game Over sprite, which will stop all the scripts.

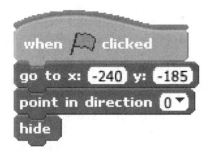

```
when  clicked
forever
    if   touching alien1 ? or  touching alien2 ?
        switch to costume blast
        repeat 10
            change whirl effect by 25
        clear graphic effects
        hide
        broadcast GameOver
```

Move on to the bullet sprite and add these code blocks to move it into place. The bullet will be hidden until the space bar is pressed:

```
when  clicked
go to x: -240 y: -185
point in direction 0
hide
```

Each time the space bar is pressed, the bullet will be moved to the cannon's current location, made visible, and then will move toward the top of the screen until it touches top edge.

Here are the blocks to handle the space bar:

```
when [flag] clicked
forever
    if < key [space ▼] pressed? >
        go to x: ([x position ▼] of [cannon ▼])  y: ([y position ▼] of [cannon ▼])
        show
        repeat until < touching [edge ▼] ? >
            move (10) steps
        hide
```

The final sprite is the Game Over message. It's role is simple; hide until it receives the *GameOver* message, at which point it appears and stops all the scripts:

```
when [flag] clicked
go to x: (0) y: (0)
hide
```

```
when I receive [GameOver ▼]
show
stop all
```

That's the end of the game; if you want to see all the blocks together, see Appendix B.

Scratch and the Real World

If you poke around the sensing palette you'll notice two interesting blocks: *sensor* and *sensor value*. These blocks are used to interact with an external sensor accessory called the PicoBoard (see Figure 5-6). The PicoBoard has a microcontroller on board that reads the sensors and sends the values to Scratch over a USB connection.

The PicoBoard provides a button, a slider, a sound sensor, and jacks for four arbitrary analog inputs. Special alligator clip connectors can be used to take analog samples from many different sources.

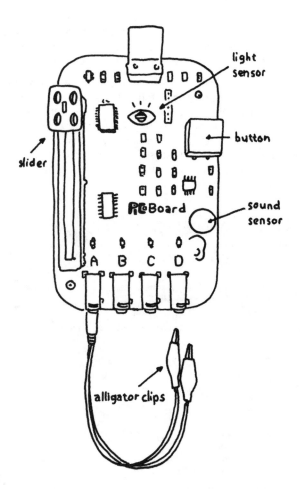

Figure 5-6. *The PicoBoard is a sensor accessory that is designed to work with Scratch.*

The PicoBoard send the sensor values to Scratch using its own PicoBoard protocol. The S4A (*http://seaside.citilab.eu/scratch/arduino*) Project (Scratch for Arduino) has implemented the same protocol for Arduino, so you can connect to the real world that way as well. S4A requires a custom version of Scratch that's available on their project page.

Sharing Your Programs

One of the really interesting aspects of Scratch is that there's a community built right into the program. Besides *Random Sprite* tool (which will grab random sprites from other Scratch users), there's also a sharing feature that will let you package and upload your programs to MIT's Scratch project page (*http://scratch.mit.edu/*). At last count there were over 1.2 million Scratch users, who have shared over 2.8 million programs.

One reason there are so many shared projects is that it is very easy to do so. Once you create an account at scratch.mit.edu (*http://scratch.mit.edu/*), just select *Share This Project Online...* under the Share menu. You'll be prompted for some information (Figure 5-7) and your project will be uploaded to the site. There's a 10MB size limit, so you might need to compress some of your images or sounds first (see the options under the Edit menu). The Scratch project page is a great place to go to see what is possible with the environment.

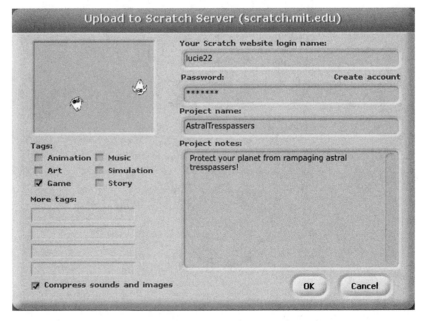

Figure 5-7. *When you've got something good, share it on the MIT Scratch project site using the built-in sharing tool. Unfortunately, the web site uses Flash so you won't be able to use it directly from the Raspberry Pi.*

Going Further

The main Scratch wiki (http://wiki.scratch.mit.edu/wiki/Scratch_Wiki)
    MIT's repository for Scratch referene material.

MIT's Scratch page (http://scratch.mit.edu/)
    This is the Scratch community site: literally millions of projects and registered members.

6/Arduino and the Pi

As you'll see in the next few chapters, you can use the GPIO pins on the Raspberry Pi to connect to sensors or things like blinking LEDs and motors. If you have experience using the Arduino microcontroller development platform, you can also use that alongside with the Raspberry Pi.

When the Raspberry Pi was first announced, a lot of people asked if this was an Arduino-killer. For about the same price you can get much more processing power; why use Arduino when you have a Pi? It turns out the two platforms are actually complementary, and the Raspberry Pi makes a great host for the Arduino. There are a few other cases where you might want to Arduino and the Pi together:

- There are lots of libraries and sharable examples for the Arduino.
- If you already have a good working Arduino project that you want to supplement with more processing power. For example, maybe you have a MIDI controller that was hooked up to a synthesizer, but now you want to upgrade to synthesizing the sound directly on the Pi.
- When you're dealing with 5V logic levels. The Pi operates at 3.3V, and its pins are not tolerant of 5V.
- You may be prototyping something a little out of your comfort zone and may make some chip-damaging mistakes. I've seen students try to drive motors directly from a pin on the Arduino (don't try it); it was easy to pry out the damaged microcontroller chip out of its socket and replace it (less than $10 usually). No so with the Raspberry Pi.
- When you have a problem that requires exact control in real time, such as a controller for a 3D printer. As we saw in Chapter 3 Raspbian is not a Real Time operating system, and programs can't necessarily depend on the same "instruction per clock cycles" rigor of a microcontroller.

The examples in this section assume that you know at least the basics of using the Arduino development board and Integrated Development Environment (IDE). If you don't have a good grasp of the fundamentals, Getting

Started with Arduino (*http://shop.oreilly.com/product/0636920021414.do*) by Massimo Banzi is a great place to start. The official Arduino tutorials (*http://arduino.cc/en/Tutorial/HomePage*) are quite good as well, and provide a lot of opportunities to cut and paste good working code.

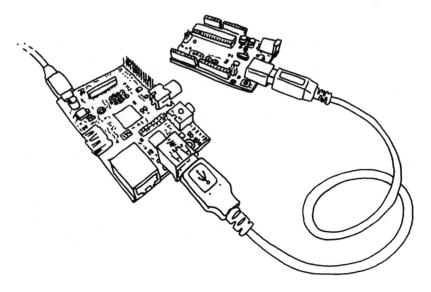

Figure 6-1. *Arduino and the Raspberry Pi are BFFs.*

Installing Arduino in Raspbian

To program an Arduino development board you need to hook it up to a computer with a USB cable, then compile and flash a program to the board using the Arduino IDE. You can do this with any computer, or you can use your Raspberry Pi as a host to program the Arduino.

Using the Raspberry Pi to program the Arduino will be quicker to debug, but compiling will be a little slower on the Pi than on a modern laptop or desktop computer. It's not too bad though, and you'll find that compiling will take less time after the very first compile, as Arduino only compiles code that has changed since the last compilation.

To install the Arduino IDE on the Raspberry Pi:

```
sudo apt-get update ❶
sudo apt-get install arduino ❷
```

❶   Make sure you have the latest package list

❷   Download the Arduino package

This command will install Java plus a lot of other dependencies. The Arduino environment will appear under the *Electronics* section of the program menu (don't launch it just yet though).

If you're running the Pi headless you can just plug the Arduino into one of the open USB ports. If you don't have an open USB port, you may be able to use a free port on your keyboard, or else you'll need a USB hub. The USB connection should be able to provide enough power for the Arduino, but you might want to power the Arduino separately for good measure.

 Note that you'll need to plug the Arduino USB cable in after the Raspberry Pi has booted up. If you leave it plugged in at boot time the Raspberry Pi may hang as it tries to figure out all the devices on the USB bus.

When you launch the Arduino IDE, it polls all the USB devices and builds a list that is shown in the Tools→Serial Port menu. In order to access the serial port you'll need to make sure that the *pi* user has permission to do so. You can do that by adding the *pi* user to the *tty* and *dialout* groups. You'll need to do this before running the Arduino IDE.

```
sudo usermod❶ -a -G❷ tty pi
sudo usermod -a -G dialout pi
```

❶   usermod is a Linux program to manage users

❷   -a -G puts the user (pi) in the specified group (tty, then dialout)

Now you can run Arduino. Click Tools→Serial Port and select the serial port (most likely */dev/ttyACM0*), then click Tools→Board select the type of Arduino Board you have (e.g. *Uno*). Click File→Examples→01.Basics→Blink to load a basic example sketch. Click the Upload button in the toolbar or choose File→Upload to upload the sketch, and after the sketch loads, the Arduino light will start blinking.

Finding the Serial Port

If, for some reason, */dev/ttyACM0* doesn't work, you'll need to do a little detective work. To find the USB serial port that the Arduino is plugged into without looking at the menu, try the following from the command line. Without the Arduino connected, type:

```
ls /dev/tty*
```

Plug in Arduino, then try the same command again and see what changed. On my Raspberry Pi, at first I had **/dev/ttyAMA0** listed (which is the onboard USB hub). When I plugged in the Arduino **/dev/ttyACM0** popped up in the listing.

Improving the User Experience

While you're getting set up, you may notice that that quality of the default font in the Arduino editor is less than ideal. You can improve it by downloading the open source font Inconsolata. To install, type:

```
sudo apt-get install ttf-inconsolata
```

Then edit the Arduino preferences file:

```
nano ~/.arduino/preferences.txt
```

and change the following lines to:

```
editor.font=Inconsolata,medium,14
editor.antialias=true
```

When you restart Arduino, the editor will use the new font.

Talking in Serial

To communicate between the Raspberry Pi and the Arduino over a serial connection, you'll use the built-in *Serial* library on the Arduino side, and the Python pySerial (*http://pyserial.sourceforge.net/*) module on the Raspberry Pi side. To install the serial module:

```
sudo apt-get install python-serial python3-serial
```

Open the Arduino IDE and upload this code to the Arduino:

```
void setup() {
  Serial.begin(9600);
}

void loop() {
  for (byte n = 0; n < 255; n++) {
   Serial.write(n);
   delay(50);
  }
}
```

This counts upward and sends each number over the serial connection.

Note that in Arduino `Serial.write()` sends the actual number; the string "123" instead of the number 123, use the `Serial.print()` command.

Next you'll need to know which USB serial port the Arduino is connected to (see "Finding the Serial Port" (page 79)). Here's the Python script; if the port isn't */dev/ttyACM0*, change the value of **port**. (See Chapter 3 for more on Python). Save it as *SerialEcho.py* and run it with `python SerialEcho.py`:

```
import serial

port = "/dev/ttyACM0"
serialFromArduino = serial.Serial(port,9600) ❶
serialFromArduino.flushInput() ❷
while True:
    if (serialFromArduino.inWaiting() > 0):
        input = serialFromArduino.read(1) ❸
        print(ord(input)) ❹
```

❶ Open the serial port connected to the Arduino.

❷ Clear out the input buffer.

❸ Read one byte from the serial buffer.

❹ Change the incoming byte into an actual number with `ord()`.

You won't be able to upload to Arduino when Python has the serial port open, so make sure you kill the Python program with `Control-C` before you upload the sketch again. You will be able to upload to an Arduino Leonardo or Arduino Micro, but doing so will break the connection with the Python script, so you'll need to restart it anyhow.

The Arduino is sending a number to the Python script, which interprets that number as a string. The **input** variable will contain whatever character maps to that number in the ASCII table (*http://en.wikipedia.org/wiki/ASCII*). To get a better idea, try replacing the last line of the Python script with this:

```
print(str(ord(input)) + " = the ASCII character " + input + ".")
```

Setting the Serial Port as an Argument

If you want to set the port as a command line argument, use the sys module to grab the first argument:

```
import serial, sys

if (len(sys.argv) != 2):
    print("Usage: python ReadSerial.py port")
    sys.exit()
port = sys.argv[1]
```

After you do this, you can run the program like this: python SerialE cho.py /dev/ttyACM0.

The first simple example just sent a single byte; this could be fine if you are just sending a series of event codes from the Arduino. For example, if the left button is pushed, send a 1, if the right send 2. That's only good for 255 discrete events, though; more often you'll want to send arbitrarily large numbers or strings. If you're reading analog sensors with the Arduino, for example, you'll want to send numbers in the range 0 to 1023.

Parsing arbitrary numbers that come in one byte at a time can be trickier than you might think in many languages. The way Python and PySerial handles strings makes it almost trivial, however. As a simple example, update your Arduino with the following code that counts from 0 to 1024.

```
void setup() {
  Serial.begin(9600);
}

void loop() {
  for (int n = 0; n < 1024; n++)
    Serial.println(n, DEC);
    delay(50);
  }
}
```

The key difference is in the println() command. In the previous example, the Serial.write() function was used to write the raw number to the serial port. With println(), the Arduino formats the number as a decimal string and sends the ASCII codes for the string. So instead of sending the number 254, it sends the string "254\r\n". The \r represents a carriage return and the \n represents a newline (these are concepts that carried over from the type-writer into computing: carriage return moves to the start of the line, newline starts a new line of text).

On the Python side, you can use `readline()` instead of `read()`, which will read all of the characters up until (and including) the carriage return and newline. Python has a flexible set of functions for converting between the various data types and strings. It turns out you can just use the `int()` function to change the formatted string into an integer:

```python
import serial

port = "/dev/ttyACM0"
serialFromArduino = serial.Serial(port,9600)
serialFromArduino.flushInput()
while True:
    input = serialFromArduino.readline() ❶
    inputAsInteger = int(input) ❷
    print(inputAsInteger * 10) ❸
```

❶ Read the whole string into the `input` variable.

❷ Convert to an integer.

❸ Print it, but multiply by 10 first to prove that it is really an integer and not a string.

Note that it is simple to adapt this example so that it will read an analog inputs and send the result; just change the loop to:

```c
void setup() {
  Serial.begin(9600);
}

void loop() {
  int n = analogRead(A0);
  Serial.println(n, DEC);
  delay(100);
}
```

Assuming you change the Python script to just print `inputAsInteger` instead of `inputAsInteger * 10`, you should get some floating values in the 200 range if nothing is connected to analog pin 0. With some jumper wire, connect the pin to GND and the value should be 0. Connect it to the 3V3 pin and you'll see a value around 715, and 1023 when connected to the 5V pin.

Going Further

You'll find that the setup is almost trivial once you have the right software, and many projects will look the same as far as the code for basic communication goes. As with any form of communication, things get tricky once you get past "hello world"; you'll need to create *protocols* (or find an existing protocol and implement it) so that each side understands the other. The details of serial protocols is beyond the scope of this book, but there are a lot of great examples of how other people have solved problems in the Interfacing with Software (*http://www.arduino.cc/playground/Main/InterfacingWith Software*) section of the Arduino Playground (*http://www.arduino.cc/play ground/*).

Firmata
> Hans-Christoph Steiner's Firmata (*http://arduino.cc/en/Reference/ Firmata*) is an all-purpose serial protocol that has been around for quite a while. Firmata is simple and human readable; while it may not be great for all applications it is a good place to start.

MIDI
> If your project is musical, consider using MIDI commands as your serial protocol. Since MIDI is (basically) just serial, it should Just Work.

Arduino-compatible Raspberry Pi shields
> There are a few daughterboards (or shields) on the market that connect the GPIO pins on the Raspberry Pi with an Arduino-compatible microcontroller. WyoLum's A la Mode (*http://baldwisdom.com/projects/ alamode/*) shield is a good solution, and offers a few other accessories like a real time clock.

Talk over a network
> Finally, you can ditch the serial connection all together and use talk to the Arduino over a network. A lot of really interesting projects are using the WebSocket (*http://www.websocket.org/*) protocol along with the Node.js (*http://nodejs.org/*) Javascript platform. The Noduino project (*http://semu.github.com/noduino/*) is a good place to start exploring this technology.

Using the serial pins on the Raspberry Pi header
> The header on the Raspberry Pi pulls out a number of input and output pins, including two that can be used to send and receive serial data bypassing the USB port. To do that you'll need to first cover the material in Chapter 8, and make sure that you have a level shifter to protect the Raspberry Pi 3.3V pins from the Arduino's 5V pins.

If you're looking to get deeper into making physical devices communicate, a good starting point is Making Things Talk (*http://shop.oreilly.com/product/ 9780596510510.do*) by Tom Igoe.

7/Basic Input and Output

While the Raspberry Pi is, in essence, a very inexpensive Linux computer, there are a few things that distinguish it from laptop and desktop machines that we usually use for writing email, browsing the web, or word processing. One of the main differences is that the Raspberry Pi can be directly used in electronics projects because it has *general purpose input and output* pins right on the board, shown in Figure 7-1.

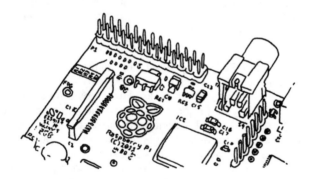

Figure 7-1. *Raspberry Pi's GPIO Pins*

These GPIO pins can be accessed for controlling hardware such as LEDs, motors, and relays, which are all examples of outputs. As for inputs, your Raspberry Pi can read the status of buttons, switches, and dials, or it can read sensors like temperature, light, motion, or proximity sensors (among many others).

One of the drawbacks to the Raspberry Pi is that there's no way to directly connect *analog sensors*, such as light and temperature sensors. Doing so requires a chip called an *analog to digital converter* or *ADC*. See Appendix C for how to read analog sensors using an ADC.

The best part of having a computer with GPIO pins is that you can create programs to read the inputs and control the outputs based on many different conditions, as easily as you'd program your desktop computer. Unlike a typical microcontroller board, which also has programmable GPIO pins, the Raspberry Pi has a few extra inputs and outputs such as your keyboard, mouse, and monitor, as well as the Ethernet port, which can act as both an input and an output. If you have experience creating electronics projects with microcontroller boards like the Arduino, you have a few more inputs and outputs at your disposal with the Raspberry Pi. Best of all, they're built right in; there's no need to wire up any extra circuitry to use them.

Having a keyboard, mouse, and monitor is not the only advantage that Raspberry Pi has over typical microcontroller boards. There are a few other key features that will help you in your electronics projects:

Filesystem
Being able to read and write data in the Linux file system will make many projects much easier. For instance, you can connect a temperature sensor to the Raspberry Pi and have it take a reading once a second. Each reading can be appended to the end of a log file, which can be easily downloaded and parsed in a graphing program. It can even be graphed right on the Raspberry Pi itself!

Linux tools
Packaged in the Raspberry Pi's Linux distribution is a set of core command-line utilities, which let you work with files, control processes, and automate many different tasks. These powerful tools are at your disposal for all of your projects. And since there is an enormous community of Linux users that depend on these core utilities, getting help is usually one web search away. For general Linux help, you can usually find answers at Stack Overflow (*http://stackoverflow.com*). If you have a question specific to Raspberry Pi, try the Raspberry Pi Forum (*http://www.raspberrypi.org/phpBB3/*) or the Raspberry Pi section of Stack Overflow (*http://raspberrypi.stackexchange.com*).

Languages
There are many programming languages out there and embedded Linux systems like the Raspberry Pi give you the flexibility to choose whichever language you're most comfortable with. The examples in this book will use shell scripting and Python but they could easily be translated to languages like C, Java, Perl, or many others.

Using Inputs and Outputs

There are a few supplies that you'll need in addition to the Raspberry Pi itself in order to try out these basic input and output tutorials. Many of these parts you'll be able to find in your local RadioShack, or they can be ordered online from stores like Maker Shed, Sparkfun, Adafruit, Mouser, or Digi-Key. Here are a few of the basic parts:

- Solderless breadboard
- LEDs, assorted
- Male-to-male jumper wires
- Female-to-male jumper wires (These are not as common as their male-to-male counterparts but are needed to connect the Raspberry Pi's GPIO pins to the breadboard.)
- Pushbutton switch
- Resistors, assorted

To make it easier to connect breadboarded components to the Raspberry Pi's pins, we also recommend Adafruit's Pi Cobbler Breakout Kit. This eliminates the need to use female-to-male jumper wires. The kit comes unassembled so it's up to you to solder the parts onto the board, but it's easy and Adafruit's tutorial (*http://learn.adafruit.com/adafruit-pi-cobbler-kit/over view*) walks you through the process step-by-step. The Pi Cobbler Breakout Kit is included, along with the other components listed above (with the exception of the female-to-male jumper wires, which are not needed if you have the Pi Cobbler Breakout Kit), in MAKE's Raspberry Pi Starter Kit (*http://oreil.ly/pikit*).

In Figure 7-2, we've labeled each pin according to its default GPIO signal number, which is how you'll refer to a particular pin in the commands you execute and in the code that you write. The unlabeled pins are assigned to other functions by default.

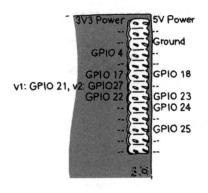

Figure 7-2. *The default GPIO pins on the Raspberry Pi. In recent revisions of the board, GPIO pin 21 was swapped for GPIO pin 27.*

You may have noticed that one of the pins has two different GPIO pin numbers in Figure 7-2. In recent versions of the board, GPIO pin 21 became GPIO pin 27. To determine the version of your board, type `cat /proc/cpuinfo` on the command line. If your revision number is listed as 0002 or 0003, you have the first version of the board. If you have a higher number, or a letter, you have a later version of the board.

Digital Output: Lighting Up an LED

The easiest way to use outputs with the GPIO pins is by connecting an LED, or light emitting diode. You can then use the Linux command line to turn the LED on and off. Once you have an understanding of how these commands work, you're one step closer to having an LED light up to indicate when you have new email, when you need to take an umbrella with you as you leave your house, or when it's time to go to bed. It's also very easy to go beyond a basic LED and use a relay to control a lamp on a set schedule, for instance.

Beginner's Guide to Breadboarding

If you've never used a breadboard (Figure 7-3) before, it's important to know which terminals are connected together. In the diagram below, we've shaded the terminal connections on a typical breadboard. Note that the power buses on the left side are not connected to the power buses on the right side of the breadboard. You'll have to use male-to-male jumper cables to connect them to each other if you need ground and voltage on both sides of the breadboard.

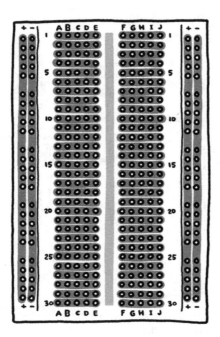

Figure 7-3. *Breadboard*

1. Using a male-to-female jumper wire, connect pin 25 on the Raspberry Pi to the breadboard. Refer to Figure 7-2 for the location of each pin on the Raspberry Pi's GPIO header.

2. Using another jumper wire, connect the Raspberry Pi's ground pin to the negative power bus on the breadboard.

3. Now you're ready to connect the LED (see Figure 7-4). Before you do that, it's important to know that LEDs are *polarized*: it matters which of the LED's wires is connected to what. Of the two leads coming off the LED, the longer one is the anode and should be connected to a GPIO pin. The shorter lead is the cathode and should be connected to ground. Another way to tell the difference is by looking from the top. The flat side of the LED indicates the cathode, the side that should be connected to ground. Insert the anode side of the LED into the breadboard in the same channel as the jumper wire from pin 25, which will connect pin 25 to the LED. Insert the cathode side of the LED into the ground power bus.

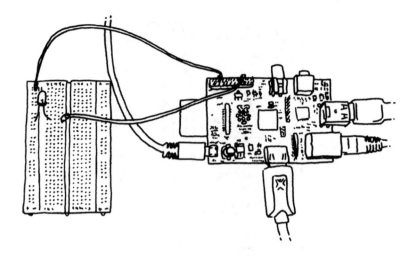

Figure 7-4. *Connecting an LED to the Raspberry Pi*

1. With your keyboard, mouse, and monitor hooked up, power on your Raspberry Pi and log in. If you're at a command line, you're ready to go. If you're in the X Window environment, double click on the LXTerminal icon on your desktop. This will bring up a terminal window.

2. In order to access the input and output pins from the command line, you'll need to run the commands as root, the *superuser* account on the Raspberry Pi. To start running commands as root, type **sudo su** at the command line and press enter:

   ```
   pi@raspberrypi ~ $ sudo su
   root@raspberrypi:/home/pi#
   ```

   You'll notice that the command prompt has changed, indicating that you're now running commands as root.

 The root account has administrative access to all the functions and files on the system and there is very little protecting you from damaging the operating system if you type a command that can harm it, so exercise caution when running commands as root. If you do mess something up, don't worry about it too much; you can always reimage the SD card with a clean Linux install. When you're done working within the root account, type exit to return to working within the pi user account.

---

6. Before you can use the command line to turn the LED on pin 25 on and off, you need to *export the pin to the userspace* (in other words, make the pin available for use outside of the confines of the Linux kernel), this way:

```
root@raspberrypi:/home/pi# echo 25 > /sys/class/gpio/export
```

The echo command writes the number of the pin you want to use (25) to the export file, which is located in the folder /sys/class/gpio. When you write pin numbers to this special file, it creates a new directory in /sys/class/gpio that has the control files for the pin. In this case, it created a new directory called /sys/class/gpio/gpio25.

7. Change to that directory with the cd command and list the contents of it with ls:

```
root@raspberrypi:/home/pi# cd /sys/class/gpio/gpio25
root@raspberrypi:/sys/class/gpio/gpio25# ls
active_low  direction  edge  power  subsystem  uevent  value
```

The command cd stands for "change directory." It changes the working directory so that you don't have to type the full path for every file. ls will list the files and folders within that directory There are two files that you're going to work with in this directory: direction and value.

8. The direction file is how you'll set this pin to be an input (like a button) or an output (like an LED). Since you have an LED connected to pin 25 and you want to control it, you're going to set this pin as an output:

```
root@raspberrypi:/sys/class/gpio/gpio25# echo out > direction
```

9. To turn the LED on, you'll use the echo command again to write the number 1 to the value file:

```
root@raspberrypi:/sys/class/gpio/gpio25# echo 1 > value
```

10. After pressing enter, the LED will turn on! Turning it off is as simple as using echo to write a zero to the value file:

```
root@raspberrypi:/sys/class/gpio/gpio25# echo 0 > value
```

Virtual Files

The files that you're working with aren't actually files on the Raspberry Pi's SD card, but rather are a part of Linux's *virtual file system*, which is a system that makes it easier to access low-level functions of the board in a simpler way. For example, you could turn the LED on and off by writing to a particular section of the Raspberry Pi's memory, but doing so would require more coding and more caution.

So if writing to a file is how you control components that are outputs, how do you check the status of components that are inputs? If you guessed "reading a file," then you're absolutely right. Let's try that now.

Digital Input: Reading a Button

Simple pushbutton switches like the one in Figure 7-5 are great for controlling basic digital input and best of all, they're made to fit perfectly into a breadboard.

 These small buttons are very commonly used in electronics projects and understanding what's going on inside of them will help you as you prototype your project. When looking at the button as it sits in the breadboard (see Figure 7-5): the top two terminals are always connected to each other. The same is true for the bottom two terminals; they're always connected. When you push down on the button, these two sets of terminals are connected to each other.

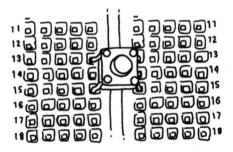

Figure 7-5. *Button*

When you read a digital input on a Raspberry Pi, you're checking to see if the pin is connected to either 3.3 volts or to ground. It's important to remember that it must be either one or the other, and if you try to read a pin that's not connected to either 3.3 volts or ground, you'll get unexpected results. Once you understand how digital input with a pushbutton works, you can start using components like magnetic security switches, arcade joysticks, or even vending machine coin accepters.

1. Insert the pushbutton into the breadboard so that its leads straddle the middle channel.

2. Using a jumper wire, connect pin 24 from the Raspberry Pi to one of the top terminals of the button.

3. Connect the 3V3 pin from the Raspberry Pi to the positive power bus on the breadboard.

 Be sure that you connect the button to the 3V3 pin and not the 5V pin. Using more than 3.3 volts on an input pin will permanently damage your Raspberry Pi.

4. Connect one of the bottom terminals of the button to the power bus. Now when you push down on the button, the 3.3 volts will be connected to pin 24.

5. Remember what we said about how a digital input must be connected to *either* 3.3 volts or ground? When you let go of the button, pin 24 isn't connected to either of those and is therefore *floating*. This condition will cause unexpected results, so let's fix that. Use a 10K resistor (labeled with the colored bands: brown, black, orange, and then silver or gold) to connect the input side of the switch to the ground rail, which you connected to the Raspberry Pi's ground in the output example. When the switch is not pressed, the pin will be connected to ground.

Since electricity always follows the path of least resistance towards ground, when you press the switch, the 3.3 volts will go towards the Raspberry Pi's input pin, which has less resistance than the 10K resistor. When everything's hooked up, it should look like Figure 7-6.

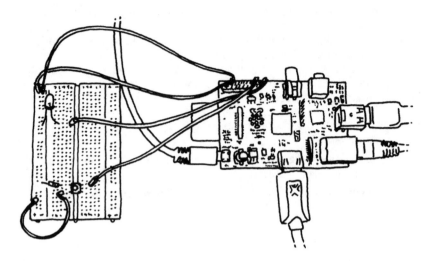

Figure 7-6. *Connecting a button to the Raspberry Pi*

6. Now that the circuit is built, let's read the value of the pin from the command line. If you're not already running commands as root, type **sudo su**.

7. As with the previous example, you need to export the input pin to userspace:

```
root@raspberrypi:/sys/class/gpio/gpio25# echo 24 > /sys/class/gpio/
export
```

8. Let's change to the directory that was created during the export operation:

```
root@raspberrypi:/sys/class/gpio/gpio25# cd /sys/class/gpio/gpio24
```

9. Now set the direction of the pin to input:

```
root@raspberrypi:/sys/class/gpio/gpio24# echo in > direction
```

10. To read the value of the of the pin, you'll use the **cat** command, which will print the contents of files to the terminal. The command **cat** gets its name because it can also be used to concatenate, or join, files together. It can also display the contents of a file for you.

```
root@raspberrypi:/sys/class/gpio/gpio24# cat value
0
```

11. The zero indicates that the pin is connected to ground. Now press and hold the button while you execute the command again:

```
root@raspberrypi:/sys/class/gpio/gpio24# cat value
1
```

12. If you see the number one, you'll know you've got it right!

 To easily execute a command that you've previously executed, hit the up key until you see the command that you want to execute and hit enter.

Now that you can use the Linux command line to control an LED or read the status of a button, let's use a few of Linux's built-in tools to create a very simple project that uses digital input and output.

Project: Cron Lamp Timer

Let's say you're leaving for a long vacation early tomorrow morning and you want to ward off would-be burglars from your home. A lamp timer is a good deterrent, but hardware stores are closed for the night and you won't have time to get one before your flight in the morning. However, since you're a Raspberry Pi hobbyist, you have a few supplies lying around, namely:

- Raspberry Pi board
- Breadboard
- Jumper wires, female-to-male.
- PowerSwitch Tail II relay
- Hookup wire

With these supplies, you can make your own programmable lamp timer using two powerful Linux tools: *shell scripts* and *cron*.

Scripting Commands

A shell script is a file that contains a series of commands (just like the ones you've been using to control and read the pins). Take a look at the shell script below and the explanation of the key lines.

```
#!/bin/bash # ❶
echo Exporting pin $1. # ❷
echo $1 > /sys/class/gpio/export # ❸
echo Setting direction to out.
echo out > /sys/class/gpio/gpio$1/direction # ❹
echo Setting pin high.
echo 1 > /sys/class/gpio/gpio$1/value
```

❶ This line is required for all shell scripts.

❷ "$1" refers to the first command line argument.

❸ Instead of exporting a specific pin number, the script uses the first command line argument.

❹ Notice that the first command line argument replaces the pin number here as well.

Save that as a text file called **on.sh** and make it executable with the chmod command:

```
root@raspberrypi:/home/pi# chmod +x on.sh
```

 You still need to be executing theses commands as root. Type **sudo su** if you're get errors like "Permission denied."

A command line argument is a way of passing information into a program or script by typing it in after name of the command. When you're writing a shell script, $1 refers to the first command line argument, $2 refers to the second, and so on. In the case of **on.sh**, you'll type in the pin number that you want to export and turn on. Instead of *hard coding* pin 25 into the shell script, it's more universal by referring to the pin that was typed in at the command line. To export pin 25 and turn it on, you can now type:

```
root@raspberrypi:/home/pi/# ./on.sh 25 ❶
Exporting pin 25.
Setting direction to out.
Setting pin high.
```

❶ The "./" before the filename indicates that you're executing the script in the directory you're in.

If you still have the LED connected to pin 25 from earlier in the chapter, it should turn on. Let's make another shell script called off.sh, which will turn the LED off. It will look like this:

```
#!/bin/bash
echo Setting pin low.
echo 0 > /sys/class/gpio/gpio$1/value
echo Unexporting pin $1
echo $1 > /sys/class/gpio/unexport
```

Now let's make it executable and run the script:

```
root@raspberrypi:/home/pi/temp# chmod +x off.sh
root@raspberrypi:/home/pi/temp# ./off.sh 25
Setting pin low.
Unexporting pin 25
```

If everything worked, the LED should have turned off.

Connecting a Lamp

Of course, a tiny little LED isn't going to give off enough light to fool burglars into thinking that you're home, so let's hook up a lamp to the Raspberry Pi.

1. Remove the LED connected to pin 25.
2. Connect two strands of hookup wire to the breadboard, one that connects to pin 25 of the Raspberry Pi and the other to the ground bus.
3. The strand of wire that connects to pin 25 should be connected to the "+in" terminal of the PowerSwitch Tail.
4. The strand of wire that connects to ground should be connected to the "-in" terminal of the PowerSwitch Tail. Compare your circuit to Figure 7-7.
5. Plug the PowerSwitch Tail into the wall and plug a lamp into the Power-Switch Tail. Be sure the lamp's switch is in the on position.
6. Now when you execute ./on.sh 25, the lamp should turn on and if you execute ./off.sh 25, the lamp should turn off!

 Inside the PowerSwitch Tail there are a few electronic components that help you control high voltage devices like a lamp or blender by using a low voltage signal such as the one from the Raspberry Pi. The "click" you hear from the PowerSwitch Tail when it's turned on or off is the relay, the core component of the circuit inside. A relay acts like a switch for the high voltage device that can be turned on or off depending on whether the low voltage control signal from the Raspberry Pi is on or off.

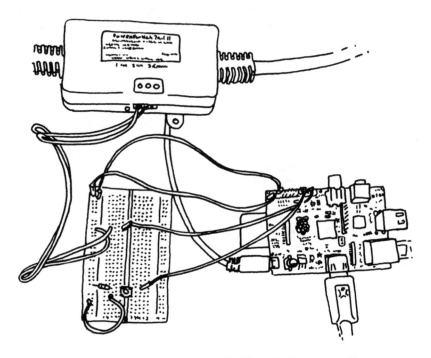

Figure 7-7. *Connecting a PowerSwitch Tail II to the Raspberry Pi*

Scheduling Commands with cron

So now you've packaged up a few different commands into two simple commands that can turn a pin on or off. And with the lamp connected to the Raspberry Pi through the PowerSwitch Tail, you can turn the lamp on or off with a single command. Now you can use cron to schedule the light to turn on and off at different times of day. cron is Linux's job scheduler. With it, you can set commands to execute on specific times and dates, or you can have jobs run on a particular period (for example, once an hour). You're going to schedule two jobs; one of them will turn the light on at 8:00pm and the other will turn the light off at 2:00 am.

 As with other time-dependent programs, you'll want to make sure you've got the correct date and time set up on your Raspberry Pi, as described in Chapter 1.

To add these jobs, you'll have to edit the cron table (a list of commands that Linux executes at specified times):

```
root@raspberrypi:/home/pi/# crontab -e
```

This will launch a text editor to change root's cron table. At the top of the file, you'll see some information about how to modify the cron table. Use your arrow keys to get to the bottom of the file and add these two entries at the end of the file.

```
0 20 * * * /home/pi/on.sh 25
0 2 * * * /home/pi/off.sh 25
```

 cron will ignore any lines that start with the hash mark. If you want to temporarily disable a line without deleting it or add a comment to the file, put a hash mark in front of the line.

Type `Control-X` to exit, type `y` to save the file when it prompts you, and hit enter to accept the default file name. When the file is saved and you're back at the command line, it should say `installing new crontab` to indicate that the changes you've made are going to be executed by `cron`.

More About Cron

Cron will let you schedule jobs for specific dates and times or on intervals. There are five time fields (or six if you want to schedule by year), each separated by a space followed by another space then the command to execute. Asterisks indicate that the job should execute each period. For example:

Table 7-1. *Cron Entry for Turning Light On at 8:00pm Every Day*

0	20	*	*	*	/home/pi/on.sh 25
Minute (: 00)	Hour (8pm)	Every Day	Every Month	Every Day of Week	path to command

Let's say you only wanted the lamp to turn on every weekday. Here's what the crontab entry would look like:

Table 7-2. *Cron Entry for Turning Light On at 8:00pm Every Weekday*

0	20	*	*	1-5	/home/pi/on.sh 25
Minute (: 00)	Hour (8pm)	Every Day	Every Month	Monday to Friday	path to command

Let's say you have a shell script that checks if you have new mail and emails you if you do. Here's how you'd get that script to run every five minutes:

Table 7-3. *Cron Entry for Checking for Mail Every Five Minutes*

*/5	*	*	*	*	/home/pi/ checkMail.sh
Every five minutes	Every Hour	Every Day	Every Month	Every Day of Week	path to command

The */5 indicates a period of every five minutes.

As you can see, cron is a powerful tool that's at your disposal for scheduling jobs for specific dates or times and scheduling jobs to happen on a specific interval.

Going Further

eLinux's Raspberry Pi GPIO Reference Page (http://elinux.org/RPi_Low-level_peripherals)
> This is the most comprehensive reference guide to the Raspberry Pi's GPIO pins.

Adafruit: MCP230xx GPIO Expander on the Raspberry Pi (http://learn.adafruit.com/mcp230xx-gpio-expander-on-the-raspberry-pi)
> If you don't have enough pins to work with, Adafruit offers this guide to using the MCP23008 chip for 8 extra GPIO pins and the MCP23017 for 16 extra GPIO pins.

8/Programming Inputs and Outputs with Python

At the end of Chapter 7, you did a little bit of programming with the Raspberry Pi's GPIO pins using a shell script. In this chapter, you're going to learn how to use Python to do the same thing... and a little more. Much like with the shell script, Python will let you access the pins by writing code to read and control the pins automatically.

The advantage that Python has over shell scripting is that the code is easier to write and is more readable. There's also a whole slew of Python modules that make it easy for you to do some complex stuff with basic code. See Table 3-2 for a list of a few modules that might be useful in your projects. Best of all, there's a Python module called raspberry-gpio-python (*http://code.google.com/p/raspberry-gpio-python/*) that makes it easy to read and control the GPIO pins. You're going to learn how to use that module in this chapter.

Installing and Testing GPIO in Python

On the most recent versions of the Raspbian distribution of Linux, the GPIO module is already installed. If you're using an older version of Raspbian, you may need to install it. To check if you have it, you'll use Python's interactive interpreter (as you learned in Chapter 3, the interactive interpreter lets you type lines of Python code to be evaluated immediately, as opposed to writing the code into a file and executing the file).

1. Go into the Python interactive interpreter as root from the terminal prompt. (Since *raspberry-gpio-python* requires root access to read and control the pins, you need to go into the Python interactive interpreter as root with the sudo command.)

```
pi@raspberrypi ~ $ sudo python
Python 2.7.3rc2 (default, May  6 2012, 20:02:25)
[GCC 4.6.3] on linux2
Type "help", "copyright", "credits" or "license" for more information.
>>>
```

2. When you're at the >>> prompt, try importing the module:

```
>>> import RPi.GPIO as GPIO
```

3. If you don't get an error, you're all set.

Otherwise, if you do get an error when you try to import the GPIO module, the installation can be done in a few simple steps thanks to apt-get, the package manager on the Raspberry Pi.

So, if you don't already have raspberry-gpio-python installed, here's how to install it:

1. Exit out of the interpreter (press Control-D or type exit() and press Return), update the apt-get package indexes, and issue the installation command for *raspberry-gpio-python*:

```
>>> exit()
pi@raspberrypi ~ $ sudo apt-get update
pi@raspberrypi ~ $ sudo apt-get install python-rpi.gpio
```

2. When that's done, go back into the Python interactive interpreter and import the module.

```
pi@raspberrypi ~ $ sudo python
Python 2.7.3rc2 (default, May  6 2012, 20:02:25)
[GCC 4.6.3] on linux2
Type "help", "copyright", "credits" or "license" for more information.
>>> import RPi.GPIO as GPIO
>>>
```

In this chapter, we'll be using Python 2.7 instead of Python 3 since one of the modules we'll be using is only installed for Python 2.x on the Raspberry Pi. When you type python at the command prompt on the Raspberry Pi, it currently runs Python 2.7 by default. This behavior could change in the future (you can run Python 2.7 explicitly by typing python2.7 instead of python).

One important difference between the two versions is how you print text to the console. You'll use print "Hello, World!" in Python 2.x but you'd use print("Hello, World!") in Python 3.

If you don't get any errors after entering the import command, you know you're ready to try it out for the first time.

1. Before you can use the pins, you must tell the GPIO module how your code will refer to them. In Chapter 7, the pin numbers we used didn't correlate to the way that they're arranged on the board. You were actually using the on-board Broadcom chip's signal name for each pin. With this Python module, you can choose to refer to the pins either way. To use the numbering from the physical layout, use GPIO.set mode(GPIO.BOARD). But let's stick with the pin numbering that you used in Chapter 7 (GPIO.setmode(GPIO.BCM)), which is what Adafruit's Pi Cobbler and similar breakout boards use for labels:

   ```
   >>> GPIO.setmode(GPIO.BCM)
   ```

2. Set the direction of pin 25 to output:

   ```
   >>> GPIO.setup(25, GPIO.OUT)
   ```

3. Connect an LED to pin 25 like you did in "Beginner's Guide to Breadboarding" (page 88).

4. Turn on the LED:

   ```
   >>> GPIO.output(25, GPIO.HIGH)
   ```

5. Turn off the LED:

   ```
   >>> GPIO.output(25, GPIO.LOW)
   ```

6. Exit the Python interactive interpreter:

   ```
   >>> exit()
   pi@raspberrypi ~ $
   ```

In Chapter 7, you learned that digital input and output signals on the Raspberry Pi must be either 3.3 volts or ground. In digital electronics, we refer to these signals as high or low respectively. Keep in mind that not all hardware out there uses 3.3 volts to indicate high; some use 1.8 volts or 5 volts. If you plan on connecting your Raspberry Pi to digital hardware through its GPIO pins, it's important that they also use 3.3 volts.

Those steps gave you a rough idea of how to control the GPIO pins by typing in Python statements directly into the interactive interpreter. Just like how you created a shell script to turn the pins on and off in Chapter 7, you're going to create a Python script to read and control the pins automatically.

Blinking an LED

To blink an LED on and off with Python, you're going to use the statements that you already tried in the interactive interpreter in addition to a few others. For the next few steps, we'll assume you're using desktop environment, but feel free to use the command line to write and execute these Python scripts if you prefer.

Figure 8-1. *Creating a new file in the home directory*

1. Open the File Manager by clicking its icon in the task bar.
2. Be sure you're in the home directory, the default being */home/pi*. If not, click on the home icon under the Places listing.

3. Create a file in your home directory called *blink.py*. Do this by right clicking in the home directory window, going to "Create New..." and then clicking "Blank File." Name the file *blink.py*.

4. Double click on *blink.py* to open it in Leafpad, the default text editor.

5. Enter the following code and save the file:

```
import RPi.GPIO as GPIO ❶
import time ❷

GPIO.setmode(GPIO.BCM) ❸
GPIO.setup(25, GPIO.OUT) ❹

while True: ❺
    GPIO.output(25, GPIO.HIGH) ❻
    time.sleep(1) ❼
    GPIO.output(25, GPIO.LOW) ❽
    time.sleep(1) ❾
```

❶ Import the code needed for GPIO control

❷ Import the code needed for for the sleep function

❸ Use the chip's signal numbers

❹ Set pin 25 as an output

❺ Create an infinite loop consisting of the indented code below it

❻ Turn the LED on

❼ Wait for one second

❽ Turn the LED off

❾ Wait for one second

 Don't forget that indentation matters in Python.

---

6. Open LXTerminal, then use these commands to make sure the working directory is your home directory, and execute the script:

```
pi@raspberrypi ~/Development $ cd ~
pi@raspberrypi ~ $ sudo python blink.py
```

7. Your LED should now be blinking!

8. Hit Ctrl+C to stop the script and return to the command line.

Try modifying the script to make the LED blink faster by using decimals in the `time.sleep()` functions. You can also try adding a few more LEDs and getting them to blink in a pattern. You can use any of the dedicated GPIO pins: 4, 17, 18, 21, 22, 23, 24, or 25 as shown in Figure 7-2.

Reading a Button

If you want something to happen when you press a button, one way to do that is to use a technique called *polling*. Polling means continually checking over and over again for some condition. In this case, it will be polling whether the button is connecting the input pin to 3.3 volts or to ground. To learn about polling, you'll create a new Python script that will display text on screen when the user pushes a button.

1. Connect a button the same way as in "Digital Input: Reading a Button" (page 92), using pin 24 as the input. Don't forget the pull-down resistor, which goes between ground and the input pin.

2. Create a file in your home directory called *button.py* and open it in the editor.

3. Enter the following code:

```
import RPi.GPIO as GPIO
import time

GPIO.setmode(GPIO.BCM)
GPIO.setup(24, GPIO.IN) ❶

count = 0 ❷

while True:
    inputValue = GPIO.input(24) ❸
    if (inputValue == True): ❹
        count = count + 1 ❺
        print("Button pressed " + str(count) + " times.") ❻
    time.sleep(.01) ❼
```

❶ Set pin 24 as an input

❷ Create a variable called count and store 0 in it.

❸ Save the value of pin 24 into inputValue

❹ Check if that value is True (when the button is pressed)

❺ If it is, increment the counter

❻ Print the text to the terminal

❼ Wait briefly, but let other programs have a chance to run by not hogging the processor's time

4. Go back to LXTerminal and execute the script:

```
pi@raspberrypi ~ $ sudo python button.py
```

5. Now press the button. If you've got everything right, you'll see a few lines of "The button has been pressed" for each time you press the button.

The code above checks for the status of the button 100 times per second, which is why you'll see more than one sentence printed (unless you have incredibly fast fingers). The Python statement `time.sleep(.01)` is what controls how often the button is checked.

But why not continually check the button? If you were to remove the `time.sleep(.01)` statement from the program, the loop would indeed run incredibly fast, so you'd know much more quickly when the button was pressed. This comes with a few drawbacks: you'd be using the processor on the board constantly, which will make it difficult for other programs to function and it would increase the Raspberry Pi's power consumption. Since *button.py* has to share resources with other programs, you have to be careful that it doesn't hog them all up.

Now add a few lines to the code to make it a little bit better at registering a single button press:

```
import RPi.GPIO as GPIO
import time

GPIO.setmode(GPIO.BCM)
GPIO.setup(24, GPIO.IN)

count = 0

while True:
    inputValue = GPIO.input(24)
    if (inputValue == True):
        count = count + 1
        print("Button pressed " + str(count) + " times.")
        time.sleep(.3) ❶
    time.sleep(.01)
```

❶   Helps a button press register only once

This additional line of code will help to ensure that each button press is registered only once. But it's not a perfect solution. Try holding down the button. Another count will be registered and displayed three times a second, even though you're holding the button down. Try pushing the button repeatedly very quickly. It doesn't register every button press because it won't recognize distinct button presses more frequently than three times a second.

These are challenges that you'll face when you're using polling to check the status of a digital input. One way to get arround these challenges is to use an *interrupt*, which is a way of setting a specified block of code to run when the hardware senses a change in the state of the pin. There is currently experimental support for interrupts in RPi.GPIO and you can read about how to use this feature in the library's documentation (*http://pypi.python.org/pypi/ RPi.GPIO*).

Project: Simple Soundboard

Now that you know how to read the inputs on the Raspberry Pi, you can use the Python module Pygame's sound functions to make a soundboard. A soundboard lets you play small sound recordings when you push its buttons. To make your own soundboard, you'll need the following in addition to your Raspberry Pi:

- 3 pushbutton switches
- Female-to-male jumper wires
- Standard jumper wires or hookup wire, cut to size
- Solderless breadboard
- 3 resistors, 10K ohm
- Computer speakers, or an HDMI monitor that has built-in speakers

You'll also need a few uncompressed sound files, in *.wav* format. For purposes of testing, there are a few sound files preloaded on the Raspberry Pi that you can use. Once you get the soundboard working, it's easy to replace those files with any sounds you want, though you may have to convert them to *.wav* from other formats. Start off by building the circuit:

1. Using a female-to-male jumper wire, connect the Raspberry Pi's ground pin to the negative rail on the breadboard.
2. Using a female-to-male jumper wire, connect the Raspberry Pi's 3.3V pin to the positive rail on the breadboard.
3. Insert the three pushbutton switches in the breadboard, all straddling the center trench.
4. Using standard jumper wires or small pieces of hookup wire, connect the positive rail of the breadboard to the top pin of each button.
5. Now add the pulldown resistors. Connect the bottom pin of each button to ground with a 10K resistor.
6. Using female-to-male jumper wires, connect each button's bottom pin (the one with the 10K resistor) to the Raspberry Pi's GPIO pins. For this project, we used pins 23, 24, and 25.

Figure 8-2 shows the completed circuit. We created this diagram with Fritzing (*http://fritzing.org*), an open source tool for creating hardware designs.

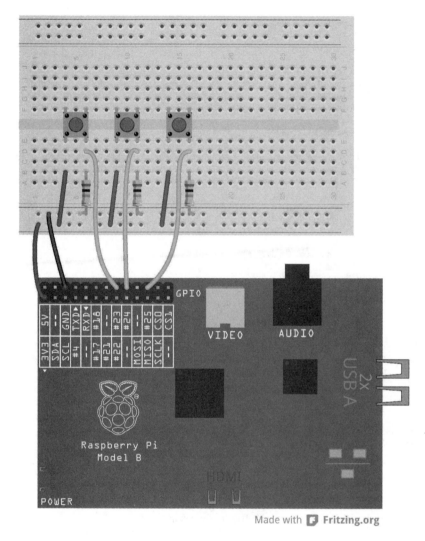

Figure 8-2. *Completed circuit for the soundboard project*

Now that you have the circuit breadboarded, it's time to work on the code.

1. Create a new directory in your home directory called *soundboard*.
2. Open that folder and create a file there called *soundboard.py*.

3. Open *soundboard.py* and type in the following code:

```
import pygame.mixer
from time import sleep
import RPi.GPIO as GPIO
from sys import exit

GPIO.setmode(GPIO.BCM)
GPIO.setup(23, GPIO.IN)
GPIO.setup(24, GPIO.IN)
GPIO.setup(25, GPIO.IN)

pygame.mixer.init(48000, -16, 1, 1024) ❶

soundA = pygame.mixer.Sound("/usr/share/sounds/alsa/Front_Center.wav") ❷
soundB = pygame.mixer.Sound("/usr/share/sounds/alsa/Front_Left.wav")
soundC = pygame.mixer.Sound("/usr/share/sounds/alsa/Front_Right.wav")

soundChannelA = pygame.mixer.Channel(1) ❸
soundChannelB = pygame.mixer.Channel(2)
soundChannelC = pygame.mixer.Channel(3)

print "Soundboard Ready." ❹

while True:
    try:
        if (GPIO.input(23) == True): ❺
            soundChannelA.play(soundA) ❻
        if (GPIO.input(24) == True):
            soundChannelB.play(soundB)
        if (GPIO.input(25) == True):
            soundChannelC.play(soundC)
        sleep(.01) ❼
    except KeyboardInterrupt: ❽
        exit()
```

❶   Initialize pygame's mixer

❷   Load the sounds

❸   Set up three channels, one for each sound so that we can play different sounds concurrently

❹   Let the user know the soundboard is ready (using Python 2 syntax)

❺   If the pin is high, execute the following line

❻   Play the sound

❼   Don't "peg" the processor by checking the buttons faster than we need to

❽   This will let us exit the script cleanly when the user hits CTRL+C, without showing the traceback message

4. Go to the command line and navigate to the folder where you've saved *soundboard.py* and execute the script with Python 2:

```
pi@raspberrypi ~/soundboard $ sudo python soundboard.py
```

5. After you see "Soundboard Ready," start pushing the buttons to play the sound samples.

 While Pygame is available for Python 3, on the Raspberry Pi's default installation, it's only installed for Python 2.

Depending on how your Raspberry Pi is set up, your sound might be sent via HDMI to your display, or it may be sent to the 3.5mm analog audio output jack on the board. To change that, exit out of the script by typing `Control+C` and executing the following command to use the analog audio output:

```
pi@raspberrypi ~/soundboard $ sudo amixer cset numid=3 1
```

To send the audio through HDMI to the monitor, use:

```
pi@raspberrypi ~/soundboard $ sudo amixer cset numid=3 2
```

Of course, the stock sounds aren't very interesting, but you can replace them with any of your own sounds: applause, laughter, buzzers, and dings. Add them to the *soundboard* directory and update the code to point to those files. If you want to use more sounds on your soundboard, add additional buttons and update the code as necessary.

Going Further

RPi.GPIO (http://code.google.com/p/raspberry-gpio-python/)
Since the RPi.GPIO library is still under active development, you may want to check the homepage for the project to find the latest updates.

Using the MCP3008 (http://learn.adafruit.com/reading-a-analog-in-and-controlling-audio-volume-with-the-raspberry-pi/overview)
Adafruit has an excellent tutorial on how to use the MCP3008 analog to digital converter to add analog sensors to your Raspberry Pi project.

9/Working with Webcams

One of the advantages to using a platform like the Raspberry Pi for DIY technology projects is that it supports a wide range of USB devices. Not only can you hook up a keyboard and mouse, but you can also connect peripherals like printers, WiFi adapters, thumb drives, additional memory cards, cameras, and hard drives. In this chapter, we're going to show a few ways that you can use a USB webcam in your Raspberry Pi projects.

While not quite as common as a keyboard and mouse, the webcam has almost become a standard peripheral for today's modern computers. Luckily for all of us, this means that a webcam from a well-known brand can be purchased for as little as $25. You can even find webcams for much less if you take a chance on an unknown brand. Best of all, many USB webcams are recognized by Linux without the need for installing additional drivers.

In Chapter 1, you may have noticed that one of the components on the board is a connector for something called the The Camera Serial Interface (CSI), pictured in Figure 9-1. Since the Broadcom chip at the core of the Raspberry Pi is meant for mobile phones and tablets, the CSI connection is how a mobile device manufacturer would connect a camera to the chip. Unfortunately, CSI cameras aren't something that consumers like us can typically buy "off the shelf" as they can with a USB webcam. At the time of press, the Raspberry Pi Foundation is developing a camera that uses the CSI connection on the board. Until that camera becomes available, we recommend using a USB webcam (Figure 9-2).

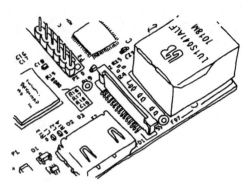

Figure 9-1. *Raspberry Pi's Camera Serial Interface*

Figure 9-2. *A typical USB webcam*

Testing Webcams

With all the different models of webcams out there, there's no guarantee that a camera will work right out of the box. If you're purchasing a webcam for use with the Raspberry Pi, search online to make sure that others have had success with the model that you're purchasing. You can also check the webcam section of eLinux.org's page of peripherals (*http://elinux.org/RPi_Verified Peripherals#USB_Webcams*) that have been verfied to work with the Raspberry Pi.

Be aware that you'll need to connect a powered USB hub to your Raspberry Pi so that you can connect your webcam in addition to your keyboard and mouse. The hub should be powered because the Raspberry Pi only lets a limited amount of electrical current through its USB ports and it's likely it will be unable to provide enough power for your keyboard, mouse, and webcam. A powered USB hub plugs into the wall and provides electrical current to the peripherals that connect to it so that they don't max out the power on your Raspberry Pi.

If you have a webcam that you're ready to test out with the Raspberry Pi, use apt-get in the terminal to install a simple camera viewing application called luvcview:

```
pi@raspberrypi ~ $ sudo apt-get install luvcview
```

After apt-get finishes the installation, run the application by typing `luvc view` in a terminal window while you're in the desktop environment. A window will open showing the view of the first video source it finds in the */dev* folder, likely */dev/video0*. Note the frame size that is printed in the terminal window. If the video seems a little choppy, you can fix this by reducing the default size of the video. For example, if the default video size is 640 by 480, close luvcview and reopen it at half the video size by typing the following at the command line:

```
pi@raspberrypi ~ $ luvcview -s 320x240
```

If you don't see video coming through, you'll want to troubleshoot here before moving on. One way to see what's wrong is by disconnecting the webcam, reconnecting it, and running the command `dmesg`, which will output diagnostic messages that might give you some clues.

Installing and Testing SimpleCV

In order to access the webcam with Python code we'll use SimpleCV (Figure 9-3), which is a feature-packed open source computer vision library. SimpleCV makes it really easy to get images from the webcam, display them on screen, or save them as files. But the best part of SimpleCV are its computer vision algorithms, which can do some pretty amazing things. Besides basic image transformations, it can also track, detect and recognize objects in an image or video. Later on this chapter, we'll try basic face detection with SimpleCV ("Face Detection" (page 122)).

Figure 9-3. *SimpleCV logo*

To install SimpleCV for Python you'll need to start by installing the other libraries it depends on. For those, you can use apt-get:

```
pi@raspberrypi ~ $ sudo apt-get install python-opencv python-scipy python-numpy python-pip
```

It's a big install and it may take a while before the process is complete. Next, you'll install the actual SimpleCV library with the following command:

```
pi@raspberrypi ~ $ sudo pip install https://github.com/ingenuitas/SimpleCV/zipball/master
```

When it's done, check that the installation worked by going into the Python interactive interpreter and importing the library:

```
pi@raspberrypi ~ $ python
Python 2.7.3rc2 (default, May  6 2012, 20:02:25)
[GCC 4.6.3] on linux2
Type "help", "copyright", "credits" or "license" for more information.
>>> import SimpleCV
>>>
```

If you get no errors after importing the library, you know you're ready to start experimenting with computer vision on the Raspberry Pi.

Displaying an Image

For many of the examples in this chapter, you'll need to work in the desktop environment so that you can display images on screen. You can work in IDLE, or save your code as *.py* files from Leafpad and execute them from the terminal window.

We're going to start you off with some SimpleCV basics using image files, and then you'll work your way up to reading images from the webcam. Once you've got images coming in from the webcam, it will be time to try some face detection.

1. Create a new directory within your home directory called *simplecv-test*.

2. Open Midori, search for an image that interests you. We decided to use a photograph of raspberries from Wikipedia.and renamed it to *raspberries.jpg*.

3. Right-click on the image and click "Save Image."

4. Save the image within the *simplecv-test* folder.

5. In the file manager (on the Accessories menu), open the *simplecv-test* folder and right-click in the folder. Choose Create New → Blank File.

6. Name the file *image-display.py*.

7. Double click on the newly created *image-display.py* file to open it in Leafpad.

8. Enter the code in Example 9-1.

9. Save the *image-display.py* file and run it from the terminal window. If you've got everything right, you'll see a photo in a new window as in Figure 9-4. You can close the window itself, or type `Control-C` in the terminal to end the script.

Figure 9-4. *The raspberry photo displayed in a window*

Example 9-1. Source code for image-display.py

```
from SimpleCV import Image, Display ❶
from time import sleep

myDisplay = Display() ❷

raspberryImage = Image("raspberries.jpg") ❸

raspberryImage.save(myDisplay) ❹

while not myDisplay.isDone(): ❺
    sleep(0.1)
```

❶   Import SimpleCV's image and display functions
❷   Creates a new window object
❸   Loads the image file *raspberries.jpg* into memory as the object image
❹   Display the image in the window
❺   Prevent the script from ending immediately after displaying the image

Modifying an Image

Now that you can load an image into memory and display it on the screen, the next step is to modify the image before displaying it. Doing this does not modify the image file itself, it simply modifies the copy of the image that's held in memory.

1. Save the *image-display.py* file as *superimpose.py*.
2. Make the enhancements to the code that are shown in Example 9-2.
3. Save the file and run it from the command line.
4. You should see the same image, but now superimposed with the shape and the text.

Example 9-2. Source code for superimpose.py

```
from SimpleCV import Image, Display, DrawingLayer, Color ❶
from time import sleep

myDisplay = Display()

raspberryImage = Image("raspberries.jpg")

myDrawingLayer = DrawingLayer((raspberryImage.width, raspberryImage.height)) ❷

myDrawingLayer.rectangle((50,20), (250, 60), filled=True) ❸
myDrawingLayer.setFontSize(45)
myDrawingLayer.text("Raspberries!", (50, 20), color=Color.WHITE) ❹
```

```
raspberryImage.addDrawingLayer(myDrawingLayer) ❺
raspberryImage.applyLayers() ❻

raspberryImage.save(myDisplay)

while not myDisplay.isDone():
    sleep(0.1)
```

❶ Import SimpleCV's drawing layer, and color functions in addition to the image and display functions you imported in the previous example

❷ Create a new drawing layer that's the same size as the image

❸ On the layer, draw a rectangle from the coordinates 50, 20 to 250, 60 and make it filled

❹ On the layer, write the text "Raspberries!" at 50, 20 in the color white

❺ Add myDrawingLayer to raspberryImage

❻ Merge the layers that have been added into raspberryImage (Figure 9-5 shows the new image)

Figure 9-5. *The modified raspberry photo*

Instead of displaying the image on screen, if you wanted to simply save your modifications to a file, Example 9-3 shows how the code would look.

Example 9-3. Source code for superimpose-save.py

```
from SimpleCV import Image, DrawingLayer, Color
from time import sleep

raspberryImage = Image("raspberries.jpg")

myDrawingLayer = DrawingLayer((raspberryImage.width, raspberryImage.height))

myDrawingLayer.rectangle((50,20), (250, 60), filled=True)
myDrawingLayer.setFontSize(45)
myDrawingLayer.text("Raspberries!", (50, 20), color=Color.WHITE)

raspberryImage.addDrawingLayer(myDrawingLayer)
raspberryImage.applyLayers()

raspberryImage.save("raspberries-titled.jpg") ❶
```

❶ Save the modified image in memory to a new filed called *raspberries-titled.jpg*

Since the code above doesn't even open up a window, you can use it from the command line without the desktop environment running. You could even modify the code to watermark batches of images with a single command.

And you're not limited to text and rectangles. Here are a few of the other drawing functions available to you with SimpleCV. Their full documentation is available at *http://simplecv.org/doc/SimpleCV.html#i/SimpleCV.DrawingLayer.DrawingLayer*

- circle
- ellipse
- line
- polygon
- bezier curve

Accessing the Webcam

Luckily, getting a webcam's video stream into SimpleCV isn't much different than accessing image files and loading them into memory. To try it out, you can make your own basic webcam viewer.

1. Create a new file named *basic-camera.py* and save the code shown in Example 9-4 in it.
2. With your webcam plugged in, run the script. You should see a window pop up with a view from the webcam as in Figure 9-6.

3. To close the window, type `Control-C` in the terminal.

Example 9-4. Source code for basic-camera.py

```
from SimpleCV import Camera, Display
from time import sleep

myCamera = Camera(prop_set={'width': 320, 'height': 240}) ❶

myDisplay = Display(resolution=(320, 240)) ❷

while not myDisplay.isDone(): ❸
    myCamera.getImage().save(myDisplay) ❹
    sleep(.1) ❺
```

❶   Create a new camera object and set the height and width of the image to 320x240 for better performance

❷   Set the size of the window to be 320x240 as well

❸   Loop the indented code below until the window is closed

❹   Get a frame from the webcam and display it in the window

❺   Wait one tenth of a second between each frame

Figure 9-6. *Outputting webcam input to the display*

You can even combine the code from the last two examples to make a Python script that will take a picture from the webcam and save it as a *.jpg* file:

```
from SimpleCV import Camera
from time import sleep
```

```
myCamera = Camera(prop_set={'width':320, 'height': 240})

frame = myCamera.getImage()
frame.save("camera-output.jpg")
```

Face Detection

One of the powerful functions that comes along with SimpleCV is called
findHaarFeatures. It's an algorithm that lets you search within an image for
patterns that match a particular profile, or *cascade*. There are a few cascades
included with SimpleCV such as face, nose, eye, mouth, and full body. Alter-
natively, you can download or generate your own cascade file if need be.
findHaarFeatures analyzes an image for matches and if it finds at least one,
the function returns the location of those matches within the image. This
means that you can detect objects like cars, animals, or people within an
image file or from the webcam. To try out findHaarFeatures, you can do some
basic face detection:

1. Create a new file in the *simplecv-test* directory called *face-detector.py*.
2. Enter the code shown in Example 9-5.
3. With your webcam plugged in and pointed at a face, run the script from
   the command line.
4. In the terminal window, you'll see the location of the faces that findHaar
   Features finds. Try moving around and watching these numbers change.
   Try holding up a photo of a face to the camera to see what happens.

Example 9-5. Source code for face-detector.py

```
from SimpleCV import Camera, Display
from time import sleep

myCamera = Camera(prop_set={'width':320, 'height': 240})

myDisplay = Display(resolution=(320, 240))

while not myDisplay.isDone():
    frame = myCamera.getImage()
    faces = frame.findHaarFeatures('face')  ❶
    if faces:  ❷
        for face in faces:  ❸
            print "Face at: " + str(face.coordinates())
    else:
        print "No faces detected."
    frame.save(myDisplay)
    sleep(.1)
```

❶   Look for faces in the image frame and save them into a faces object

❷ If findHaarFatures detected at least one face, execute the indented code below

❸ For each `face` in `faces`, execute the code below (the `print` statement) with `face` as an individual face

If your mug is on screen but you still get the message "No faces detected," try a few troubleshooting steps:

- Do you have enough light? If you're in a dark room, it may be hard for the algorithm to make out your face. Try adding more light.
- This particular Haar cascade is meant to find faces that are in their normal orientation. If you tilt your head too much or the camera isn't level, this will affect the algorithm's ability to find faces.

Project: Raspberry Pi Photobooth

You can combine different libraries together to make Python a powerful tool to do some fairly complex projects. With the GPIO library you learned about in Chapter 8 and SimpleCV, you can make your own Raspberry Pi-based photobooth that's sure to be a big hit at your next party (see Figure 9-7). And with the `findHaarFeatures` function in SimpleCV, you can enhance your photobooth with a special extra feature, the ability to automatically superimpose fun virtual props like hats, monocles, beards, and mustaches on the people in the photo booth. The code in this project is based on the Mustacheinator project in Practical Computer Vision with SimpleCV (*http://shop.oreilly.com/product/0636920024057.do*) by Kurt Demaagd, Anthony Oliver, Nathan Oostendorp, and Katherine Scott.

Figure 9-7. *Output of the Raspberry Pi Photobooth*

Here's what you'll need to turn your Raspberry Pi into a photobooth:

- A USB webcam
- A monitor
- A button, any kind you like
- Hookup wire, cut to size
- 1 resistor, 10K ohm

Before you get started, make sure that both the RPi.GPIO and SimpleCV Python libraries are installed and working properly on your Raspberry Pi. See "Installing and Testing GPIO in Python" (page 101) and "Installing and Testing SimpleCV" (page 115) for more details.

1. Like you did in Chapter 8, connect pin 24 to the button. One side of the button should be connected to 3.3V, the other to pin 24. Don't forget to use a 10K pulldown resistor between ground and the side of the switch that connects to pin 24.

2. Find or create a small image of a black mustache on a white background and save it as *mustache.png* in a new folder on your Raspberry Pi. You can also download our pre-made mustache file in the ch09 subdirectory of the downloads (see "How to Contact Us" (page xii)).

3. In that folder, create a new file called *photobooth.py*, type in the code listed in Example 9-6 and save the file.

Example 9-6. Source code for photobooth.py

```
from time import sleep, time
from SimpleCV import Camera, Image, Display
import RPi.GPIO as GPIO

myCamera = Camera(prop_set={'width':320, 'height': 240})
myDisplay = Display(resolution=(320, 240))
stache = Image("mustache.png")
stacheMask = \
    stache.createBinaryMask(color1=(0,0,0), color2=(254,254,254)) ❶
stacheMask = stacheMask.invert() ❷

GPIO.setmode(GPIO.BCM)
GPIO.setup(24, GPIO.IN)

def mustachify(frame): ❸
    faces = frame.findHaarFeatures('face')
    if faces:
        for face in faces:
            print "Face at: " + str(face.coordinates())
            myFace = face.crop() ❹
            noses = myFace.findHaarFeatures('nose')
            if noses:
```

```
                nose = noses.sortArea()[-1]  ❺
                print "Nose at: " + str(nose.coordinates())
                xmust = face.points[0][0] + nose.x - (stache.width/2)  ❻
                ymust = face.points[0][1] + nose.y + (stache.height/3)  ❼
            else:
                return frame  ❽
        frame = frame.blit(stache, pos=(xmust, ymust), mask=stacheMask)  ❾
        return frame  ❿
    else:
        return frame  ⓫

while not myDisplay.isDone():
    inputValue = GPIO.input(24)
    frame = myCamera.getImage()
    if inputValue == True:
        frame = mustachify(frame)  ⓬
        frame.save("mustache-" + str(time()) + ".jpg")  ⓭
        frame = frame.flipHorizontal()  ⓮
        frame.show()
        sleep(3)  ⓯
    else:
        frame = frame.flipHorizontal()  ⓰
        frame.save(myDisplay)
    sleep(.05)
```

❶   Create a mask of the mustache, selecting all but black to be transparent (the two parameters, color1 and color2 are the range of colors as RGB values from 0 to 255)

❷   Invert the mask so that only the black pixels in the image are displayed and all other pixels are transparent

❸   Create a function that takes in a frame and outputs a frame with a superimposed mustache if it can find the face and nose

❹   Create a subimage of the face so that searching for a nose is quicker

❺   If there are multiple nose candidates on the face, choose the largest one

❻   Set the x coordinates of the mustache

❼   Set the y coordinates of the mustache

❽   If no nose is found, just return the frame

❾   Use the blit function (short for BLock Image Transfer) to superimpose the mustache on the frame

❿   Return the "mustachified" frame

⓫   If no face is found, just return the frame

⓬   Pass the frame into the mustachify function

⓭   Save the frame as a JPG with the current time in the filename

⓮   Before showing the image, flip it horizontally so that it's a mirror image of the subject

⑮ Hold the saved image on screen for 3 seconds

⑯ If the button isn't pressed, simply flip the live image and display it.

Now you're ready to give it a try. Make sure your webcam is connected. Next, go to the terminal, and change to the directory where the mustache illustration and *photobooth.py* are and then run the script:

```
pi@raspberrypi ~ $ sudo photobooth.py
```

The output of the webcam will appear on screen. When the button is pressed, it will identify any faces, add a mustache, and save the image. (All the images will be saved in the same folder with the script).

Going Further

Practical Computer Vision with SimpleCV (http://shop.oreilly.com/product/0636920024057.do)
> This book by Kurt Demaagd, Anthony Oliver, Nathan Oostendorp, and Katherine Scott is a comprehensive guide to using SimpleCV. It includes plenty of example code and projects to further learn about working with images and computer vision in Python.

10/Python and The Internet

Python has a very active community of developers who often share their work in the form of open source libraries which simplify complex tasks. Some of these libraries make it relatively easy for us to connect our projects to the Internet to do things like get data about the weather, send an email or text message, follow trends on Twitter, or act as a web server.

In this chapter, we're going to take a look at a few ways to create Internet-connected projects with the Raspberry Pi. We'll start off by showing you how to fetch data from the Internet and then move into how you can create your own Raspberry Pi-based web server.

Download Data from a Web Server

When you type an address into your web browser and hit enter, your browser is the *client*. It establishes a connection with the *server*, which responds with a web page. Of course, a client doesn't have to be a web browser, it can also be a mail application, a weather widget on your phone or computer, or a game that uploads your high score to a global leader board. In the first part of this chapter, we're going to focus on projects that use the Raspberry Pi to act as a client. The code you'll be using will connect to internet servers to get information. Before you can do that, you'll need to install a popular Python library called Requests that is used for connecting to web servers via *hypertext transfer protocol*, or HTTP. From the terminal, here's how you install it:

```
pi@raspberrypi ~ $ sudo apt-get install python-requests
```

To confirm that Requests is installed:

```
pi@raspberrypi ~ $ python
Python 2.7.3rc2 (default, May  6 2012, 20:02:25)
```

```
[GCC 4.6.3] on linux2
Type "help", "copyright", "credits" or "license" for more information.
>>> import requests
>>>
```

If you don't get an error message, you'll know Requests has been installed properly and is now imported in this Python session. Now you can try it out:

```
>>> r = requests.get('http://www.google.com/')
>>>
```

You may be a bit disappointed at first since it seems like nothing happened. But actually, all the data from the request has been stored in the object **r**. Here's how you can display the status code:

```
>>> r.status_code
200
```

The HTTP status code 200 means that the request succeeded. There are a few other HTTP status codes in Table 10-1.

Table 10-1. *Common HTTP status codes*

Code	Meaning
200	OK
301	Moved permanently
307	Moved temporarily
401	Unauthorized
404	Not found
500	Server error

If you want to see the contents of the *response* (what the server sends back to you), try the following:

```
>>> r.text
```

If everything worked correctly, what will follow is a large block of text; you may notice some human-readable bits in there, but most of it will be hard to understand. This is the HTML of Google's landing page, which is meant to be interpreted and rendered on screen by a web browser.

However, not all HTTP requests are meant to be rendered by a web browser. Sometimes only data is transmitted, with no information about how it should be displayed. Many sites make these data protocols available to the public so that we can use them to fetch data from and send data to their servers without using a web browser. Whether it's public or not, a data protocol specification is commonly called an *application programming interface* or API. API's let different pieces of software talk to each other and are popular for sending data from one site to another over the Internet.

For example, let's say you want to make a project that will sit by your door and remind you to take your umbrella with you when rain is expected that day. Instead of setting up your own weather station and figuring out how to forecast the precipitation, you can get the day's forecast from one of many weather APIs out there.

Fetching the Weather Forecast

In order to determine whether or not it will rain today, we'll show you how to use the API from Weather Underground (*http://www.wunderground.com/*).

 Keep in mind that not all APIs are created equal and you'll have to review their documentation to determine if it's the right one for your project. Also, most APIs limit the amount of requests you can make and some even charge you to use their services. Many times, the API providers have a free tier for a small amount of daily requests, which is perfect for experimentation and personal use.

1. In a web browser, go to Weather Underground's API homepage (*http://www.wunderground.com/weather/api/*) and enter in your information to sign up.

2. After you've logged into your account, click Key Settings.

3. Notice that there's a long string of characters in the API key field (Figure 10-1). This is a unique identifier for your account that you'll need to provide in each request. If you abuse their servers by sending too many requests, they can simply shut off this API key and refuse to complete any further requests until you pay for a higher tier of service.

4. Click Documentation and then Forecast so you can see what kind of data you get when you make a forecast request. At the bottom of the page is an example URL for getting the forecast for San Francisco, CA. Notice that in the URL is your API key:

```
http://api.wunderground.com/api/YourAPIkey/forecast/q/CA/
San_Francisco.json
```

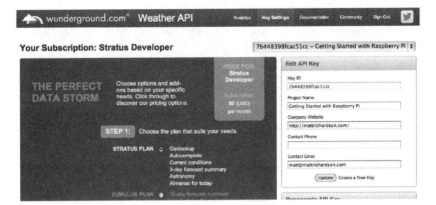

Figure 10-1. *Matt's Weather Underground API account, with the API key in the upper right.*

5. To make sure it works, go to this URL into your web browser and you should see the weather forecast data in a format called *JSON*, or JavaScript Object Notation (see Example 10-1). Notice how the data is structured hierarchically.

Even though the J stands for JavaScript, JSON is used in many programming languages, especially for communicating between applications via an API.

6. To start off, try to get today's local text forecast from within this data structure. Change the URL so that it matches your state and town and put it in a new Python script called *text-forecast.py* (Example 10-2).

7. As you see from the code, in order to get the text forecast for today, we'll have to fetch the right field from the hierarchy of the data structure (see Example 10-1). The text forecast can be found in **forecast** → **txt_fore cast** → **forecastday** → **0** (the first entry, which is today's forecast) → **fcttext**

8. When you run the script from the command line, the output should be your local weather forecast for today, such as this: "Clear. High of 47F. Winds from the NE at 5 to 10 mph."

Example 10-1. Partial JSON response from Weather Underground's API

```
"response": {
  "version": "0.1",
```

```
  "termsofService":
    "http://www.wunderground.com/weather/api/d/terms.html",
  "features": {
    "forecast": 1
  }
},
"forecast":{ ❶
  "txt_forecast": { ❷
    "date":"10:00 AM EST",
    "forecastday": [ ❸
      {
      "period":0,
      "icon":"partlycloudy",
      "icon_url":"http://icons-ak.wxug.com/i/c/k/partlycloudy.gif",
      "title":"Tuesday",
      "fcttext":
        "Partly cloudy. High of 48F. Winds from the NNE at 5 to 10 mph.", ❹
      "fcttext_metric":
        "Partly cloudy. High of 9C. Winds from the NNE at 10 to 15 km/h.",
      "pop":"0"
      },
```

❶ Forecast is the top-level parent of the data we want to access

❷ Within forecast, we want the text forecast

❸ Within the text forecast dataset, we want the daily forecasts

❹ fcttext, the text of the day's forecast

Example 10-2. Source code for text-forecast.py

```python
import requests

key = 'YOUR KEY HERE' ❶
ApiUrl = \
  'http://api.wunderground.com/api/' + key + '/forecast/q/NY/New_York.json'

r = requests.get(ApiUrl) ❷
forecast = r.json ❸
print forecast['forecast']['txt_forecast']['forecastday'][0]['fcttext'] ❹
```

❶ Replace this with your API key.

❷ Get the New York City forecast from Weather Underground (replace with your own state and city)

❸ Take the text of the JSON response and parse it into a Python dictionary object

❹ "Reach into" the data hierarchy to get today's forecast text.

So you now have a Python script that will output the text forecast whenever you need it. But how can your Raspberry Pi determine whether or not it's supposed to rain today? On one hand, you could parse the forecast text to search for words like "rain," "drizzle," "thunderstorms," "showers," and so on, but there's actually a better way. One of the fields in the data from Weather Underground API is labeled **pop**, which stands for *probability of precipitation*. With values ranging from 0 to 100%, it will give you a sense of how likely it will be raining or snowing.

For a rain forecast indicator, let's say that any probability of precipitation value over 30% is a day that we want to have an umbrella handy.

1. Connect an LED to pin 25 as you did in Figure 7-4.
2. Create a new file called *umbrella-indicator.py* and use the code in Example 10-3. Don't forget to put in your own API key and the location in the Weather Underground API URL.
3. Run the script as root with the command `sudo python umbrella-indicator.py`.

Example 10-3. Source code for umbrella-indicator.py

```
import requests
import RPi.GPIO as GPIO
import time

GPIO.setmode(GPIO.BCM)
GPIO.setup(25, GPIO.OUT)

key = 'YOUR KEY HERE' ❶
ApiUrl = \
    'http://api.wunderground.com/api/' + key + '/forecast/q/NY/New_York.json'

while True:
    r = requests.get(ApiUrl)
    forecast = r.json
    popValue = forecast['forecast']['txt_forecast']['forecastday'][0]['pop'] ❷
    popValue = int(popValue) ❸

    if popValue >= 30: ❹
        GPIO.output(25, GPIO.HIGH)
    else: ❺
        GPIO.output(25, GPIO.LOW)

    time.sleep(180) # 3 minutes ❻
```

❶   As before, change this to your API key

❷   Get today's probability of precipitation and store it in popValue

❸   Convert popValue from a string into an integer so that we can evaluate it as a number

❹   If the value is greater than 30, then turn the LED on

❺   Otherwise, turn the LED off

❻   Wait three minutes before checking again so that the script stays within the API limit of 500 requests per day.

Press Control-C to quit the program when you're done.

The Weather Underground API is one of a plethora of different APIs that you can experiment with. Table 10-2 lists a few other sites and services that have APIs.

Table 10-2. *Popular Application Programming Interfaces*

Site	API Reference URL
Facebook	*https://developers.facebook.com/*
Flickr	*http://www.flickr.com/services/api/*
Foursquare	*https://developer.foursquare.com/*
Reddit	*https://github.com/reddit/reddit/wiki/API*
Twilio	*http://www.twilio.com/*
Twitter	*https://dev.twitter.com/*
YouTube	*https://developers.google.com/youtube/*

Serving Pi (Be a Web Server)

Not only can you use the Raspberry Pi to get data from servers via the internet, but your Pi can also act as a server itself. There are many different web servers that you can install on the Raspberry Pi. Traditional web servers, like Apache or lighttpd, serve the files from your board to clients. Most of the time, servers like these are sending HTML files and images to make web pages, but they can also serve sound, video, executable programs, and much more.

However, there's a new breed of tools that extend programming languages like Python, Ruby, and JavaScript to create web servers that dynamically generate the HTML when they receive HTTP requests from a web browser. This is a great way to trigger physical events, store data, or check the value of a sensor remotely via a web browser. You can even create your own JSON API for an electronics project!

Flask Basics

We're going to use a Python web framework called Flask (*http:// flask.pocoo.org/*) to turn the Raspberry Pi into a dynamic web server. While there's a lot you can do with Flask "out of the box," it also supports many different extensions for doing things such as user authentication, generating forms, and using databases. You also have access to the wide variety of standard Python libraries that are available to you.

In order to install Flask, you'll need to have **pip** installed. If you haven't already installed **pip**, it's easy to do:

```
pi@raspberrypi ~ $ sudo apt-get install python-pip
```

After **pip** is installed, you can use it to install Flask and its dependencies:

```
pi@raspberrypi ~ $ sudo pip install flask
```

To test the installation, create a new file called *hello-flask.py* with the code from Example 10-4. Don't worry if it looks a bit overwhelming at first, you don't need to understand what every line of code means right now. The block of code that's most important is the one that contains the string "Hello World!"

Example 10-4. Source code for hello-flask.py

```
from flask import Flask
app = Flask(__name__) ❶

@app.route("/") ❷
def hello():
    return "Hello World!" ❸

if __name__ == "__main__": ❹
    app.run(host='0.0.0.0', port=80, debug=True) ❺
```

❶   Create a Flask object called **app**

❷   Run the code below when someone accesses the root URL of the server

❸   Send the text "Hello World!" to the client

❹   If this script was run directly from the command line

❺   Have the server listen on port 80 and report any errors.

Before you run the script, you need to know your Raspberry Pi's IP address (see "The Network" (page 30)). An alternative is to install avahi-daemon (run sudo apt-get install avahi-daemon from the command line). This lets you access the Pi on your local network through the address *http://raspberry pi.local*. If you're accessing the Raspberry Pi web server from a Windows machine, you may need to put Bonjour Services (*http://support.apple.com/kb/DL999*) on it for this to work.

Now you're ready to run the server, which you'll have to do as root:

```
pi@raspberrypi ~ $ sudo python hello-flask.py
 * Running on http://0.0.0.0:80/
 * Restarting with reloader
```

From another computer on the same network as the Raspberry Pi, type your Raspberry Pi's IP address into a web browser. If your browser displays "Hello World!", you know you've got it configured correctly. You may also notice that a few lines appear in the terminal of the Raspberry Pi:

```
10.0.1.100 - - [19/Nov/2012 00:31:31] "GET / HTTP/1.1" 200 -
10.0.1.100 - - [19/Nov/2012 00:31:31] "GET /favicon.ico HTTP/1.1" 404 -
```

The first line shows that the web browser requested the root URL and our server returned HTTP status code 200 for "OK." The second line is a request that many web browsers send automatically to get a small icon called a *favicon* to display next to the URL in the browser's address bar. Our server doesn't have a *favicon.ico* file, so it returned HTTP status code 404 to indicate that the URL was not found.

If you want to send the browser a site formatted in proper HTML, it doesn't make a lot of sense to put all the HTML into your Python script. Flask uses a template engine called Jinja2 (*http://jinja.pocoo.org/docs/templates/*) so that you can use separate HTML files with placeholders for spots where you want dynamic data to be inserted.

If you've still got *hello-flask.py* running, press Control-C to kill it.

To try that out, create a new file called *hello-template.py* with the code from Example 10-5. In the same directory with *hello-template.py*, create a subdirectory called *templates*. In the *templates* subdirectory, create a file called *main.html* and insert the code from Example 10-6. Anything in double curly braces within the HTML template is interpreted as a variable that would be passed to it from the Python script via the render_template function.

Example 10-5. Source code for hello-template.py

```python
from flask import Flask, render_template
import datetime
app = Flask(__name__)

@app.route("/")
def hello():
    now = datetime.datetime.now()  ❶
    timeString = now.strftime("%Y-%m-%d %H:%M")  ❷
    templateData = {
        'title' : 'HELLO!',
        'time': timeString
        }  ❸
    return render_template('main.html', **templateData)  ❹

if __name__ == "__main__":
    app.run(host='0.0.0.0', port=80, debug=True)
```

❶   Get the current time and store it in now

❷   Create a formatted string using the date and time from the now object

❸   Create a *dictionary* of variables (a set of *keys*, such as `title` that are associated with values, such as `HELLO!`) to pass into the template

❹   Return the *main.html* template to the web browser using the variables in the `templateData` dictionary

Example 10-6. Source code for templates/main.html

```html
<!DOCTYPE html>
    <head>
        <title>{{ title }}</title>  ❶
    </head>

    <body>
        <h1>Hello, World!</h1>
        <h2>The date and time on the server is: {{ time }}</h2>  ❷
    </body>
</html>
```

❶   Use the `title` variable in the HTML title of the site.

❷   Use the `time` variable on the page.

Now, when you run *hello-template.py* (as before, you need to use **sudo** to run it) and pull up your Raspberry Pi's address in your web browser, you should see a formatted HTML page with the title "HELLO!" and the Raspberry Pi's current date and time.

While it's dependent on how your network is set up, it's unlikely that this page is accessible from outside your local network via the Internet. If you'd like to make the page available from outside your local network, you'll need to configure your router for port forwarding. Refer to your router's documentation for more information about how to do this.

Connecting the Web to the Real World

You can use Flask with other Python libraries to bring additional functionality to your site. For example, with the RPi.GPIO Python module (see Chapter 8), you can create a website that interfaces with the physical world. To try it out, hook up a three buttons or switches to pins 23, 24, and 25 in the same way as the Simple Soundboard project in Figure 8-2.

The following code expands the functionality of *hello-template.py*, so copy it to a new file called *hello-gpio.py*. Add the RPi.GPIO module and a new *route* for reading the buttons, as we've done in Modified source code for hello-gpio.py. The new route will take a variable from the requested URL and use that to determine which pin to read.

You'll also need to create a new template called *pin.html*. It's not very different from *main.html*, so you may want to copy *main.html* to *pin.html* and make the appropriate changes as in Example 10-7.

Modified source code for hello-gpio.py.

```
from flask import Flask, render_template
import datetime
import RPi.GPIO as GPIO
app = Flask(__name__)

GPIO.setmode(GPIO.BCM)

@app.route("/")
def hello():
    now = datetime.datetime.now()
    timeString = now.strftime("%Y-%m-%d %H:%M")
    templateData = {
        'title' : 'HELLO!',
        'time': timeString
        }
    return render_template('main.html', **templateData)

@app.route("/readPin/<pin>") ❶
def readPin(pin):
    try: ❷
        GPIO.setup(int(pin), GPIO.IN) ❸
        if GPIO.input(int(pin)) == True: ❹
            response = "Pin number " + pin + " is high!"
```

```
        else: ❺
            response = "Pin number " + pin + " is low!"
    except: ❻
        response = "There was an error reading pin " + pin + "."

    templateData = {
        'title' : 'Status of Pin' + pin,
        'response' : response
        }

    return render_template('pin.html', **templateData)


if __name__ == "__main__":
    app.run(host='0.0.0.0', port=80, debug=True)
```

❶ Add a dynamic route with pin number as a variable.

❷ If the code indented below raises an exception, run the code in the except block

❸ Take the pin number from the URL, convert it into an integer and set it as an input

❹ If there pin is high, set the response text to say that it's high

❺ Otherwise, set the response text to say that it's low

❻ If there was an error reading the pin, set the response to indicate that

Example 10-7. Source code for templates/pin.html

```html
<!DOCTYPE html>
    <head>
        <title>{{ title }}</title> ❶
    </head>

    <body>
        <h1>Pin Status</h1>
        <h2>{{ response }}</h2> ❷
    </body>
</html>
```

❶ Insert the title provided from *hello-gpio.py* into the page's title

❷ Place the response from *hello-gpio.py* on the page inside HTML heading tags

With the above script running, when you point your web browser to your Raspberry Pi's IP address, you should see the standard "Hello World!" page we created before. But add `/readPin/24` to the end of the URL, so that it looks something like `http://10.0.1.103/readPin/24`. A page should display showing that the pin is being read as low. Now hold down the button connected to pin 24 and refresh the page; it should now show up as high!

Try the other buttons as well by changing the URL. The great part about this code is that we only had to write the function to read the pin once and create the HTML page once, but it's almost as though there are separate webpages for each of the pins!

Project: WebLamp

In Chapter 7, we showed you how to use Raspberry Pi as a simple AC outlet timer in "Project: Cron Lamp Timer" (page 95). Now that you know how to use Python and Flask, you can now control the state of a lamp over the web. This basic project is simply a starting point for creating Internet-connected devices with the Raspberry Pi.

And just like how the previous Flask example showed how you can have the same code work on multiple pins, you'll set up this project so that if you want to control more devices in the future, it's easy to add.

1. The hardware setup for this project is exactly the same as the "Project: Cron Lamp Timer" (page 95) so all the parts you need are listed there.

2. Connect the PowerSwitch Tail II relay to the pin 25, just as you did with in the Cron Lamp Timer project.

3. If you have another PowerSwitch Tail II relay, connect it to pin 24 to control a second AC device. Otherwise, just connect an LED to pin 24. We're simply using it to demonstrate how multiple devices can be controlled with the same code.

4. Create a new directory in your home directory called *WebLamp*.

5. In *WebLamp*, create a file called *weblamp.py* and put in the code from Example 10-8

6. Create a new directory within *WebLamp* called *templates*.

7. Inside *templates*, create the file *main.html*. The source code of this file can be found in Example 10-9

In terminal, navigate to the *WebLamp* directory and start the server (be sure to use Control-C to kill any other Flask server you have running first):

```
pi@raspberrypi ~/WebLamp $ sudo python weblamp.py
```

Example 10-8. Source code for weblamp.py

```python
import RPi.GPIO as GPIO
from flask import Flask, render_template, request
app = Flask(__name__)

GPIO.setmode(GPIO.BCM)

pins = {
    24 : {'name' : 'coffee maker', 'state' : GPIO.LOW},
    25 : {'name' : 'lamp', 'state' : GPIO.LOW}
    } ❶

for pin in pins: ❷
    GPIO.setup(pin, GPIO.OUT)
    GPIO.output(pin, GPIO.LOW)

@app.route("/")
def main():
    for pin in pins:
        pins[pin]['state'] = GPIO.input(pin) ❸
    templateData = {
      'pins' : pins ❹
      }
    return render_template('main.html', **templateData) ❺

@app.route("/<changePin>/<action>") ❻
def action(changePin, action):
    changePin = int(changePin) ❼
    deviceName = pins[changePin]['name'] ❽
    if action == "on": ❾
        GPIO.output(changePin, GPIO.HIGH) ❿
        message = "Turned " + deviceName + " on." ⓫
    if action == "off":
        GPIO.output(changePin, GPIO.LOW)
        message = "Turned " + deviceName + " off."
    if action == "toggle":
        GPIO.output(changePin, not GPIO.input(changePin)) ⓬
        message = "Toggled " + deviceName + "."

    for pin in pins:
        pins[pin]['state'] = GPIO.input(pin) ⓭

    templateData = {
      'message' : message,
      'pins' : pins
    } ⓮

    return render_template('main.html', **templateData)

if __name__ == "__main__":
    app.run(host='0.0.0.0', port=80, debug=True)
```

❶ Create a dictionary called pins to store the pin number, name, and pin state

❷ Set each pin as an output and make it low

❸ For each pin, read the pin state and store it in the pins dictionary

❹ Put the pin dictionary into the template data dictionary

❺ Pass the template data into the template *main.html* and return it to the user

❻ The function below is executed when someone requests a URL with the pin number and action in it

❼ Convert the pin from the URL into an int

❽ Get the device name for the pin being changed

❾ If the action part of the URL is "on," execute the code indented below

❿ Set the pin high

⓫ Save the status message to be passed into the template

⓬ Read the pin and set it to whatever it isn't (that is, toggle it)

⓭ For each pin, read the pin state and store it in the pins dictionary

⓮ Along with the pin dictionary, put the message into the template data dictionary

Example 10-9. Source code for templates/main.html

```
<!DOCTYPE html>
<head>
    <title>Current Status</title>
</head>

<body>
    <h1>Device Listing and Status</h1>

    {% for pin in pins %} ❶
    <p>The {{ pins[pin].name }} ❷
    {% if pins[pin].state == true %} ❸
        is currently on (<a href="/{{pin}}/off">turn off</a>)
    {% else %} ❹
        is currently off (<a href="/{{pin}}/on">turn on</a>)
    {% endif %}
    </p>
    {% endfor %}

    {% if message %} ❺
    <h2>{{ message }}</h2>
    {% endif %}

</body>
</html>
```

❶ Run through each pin in the pins dictionary

❷ Print the name of the pin

❸ If the pin is high, print that the device is on and link to URL to turn it off

❹ Otherwise, print that the device is off and link to the URL to turn it on

❺ If a message was passed into the template, print it.

Device Listing and Status

The coffee maker is currently on (turn off)

The lamp is currently off (turn on)

Figure 10-2. *The device interface, as viewed through a phone's web browser*

The best part about writing the code in this way is that you can very easily add as many devices that the hardware will support. Simply add the information about the device to the pins dictionary. When you restart the server, the new device will appear in the status list and its control URLs will work automatically.

There's another great feature built in: if you want to be able to flip the switch on a device with a single tap on your phone, you can create a bookmark to the address *http://ipaddress/pin/toggle*. That URL will check the pin's current state and switch it.

Going Further

Requests (http://docs.python-requests.org/en/latest/)
> The homepage for Requests includes very comprehensive documentation complete with easy to understand examples.

Flask (http://flask.pocoo.org/)
> There's a lot more to Flask that we didn't cover. The official site outlines Flask's full feature set.

Flask Extensions (http://flask.pocoo.org/extensions/)
> Flask extensions make it easy to add functionality to your site.

A/Writing an SD Card Image

While this book has focused on the Raspbian operating system, there are many other distributions that can run on the Raspberry Pi. With any of them, you need to simply download the disk image and then not-so-simply copy the disk image onto the SD card. Here's how to create an SD Card from a disk image on OS X, Windows, and Linux.

Writing an SD card from OS X

1. Open your Terminal utility (it's in */Applications/Utilities*) to get a command line prompt

2. *Without* the card in your computer's SD card reader, type `df -h`. The df program shows your free space, but it also shows which disk volumes are mounted.

3. Now insert the SD card and run `df -h` again.

4. Look at the list of mounted volumes and determine which one is the SD card by comparing it to the previous output. For example, an SD card mounts on our computer as */Volumes/Untitled*, and the device name is */dev/disk3s1*. Depending on the configuration of your computer, this name may vary. Names are assigned as devices are mounted, so you may see a higher number if you have other devices or disk images mounted in the Finder. Write the card's device name down.

5. To write to the card you'll have to unmount it first. Unmount it by typing `sudo diskutil unmount /dev/disk3s1` (using the device name you got from the previous step instead of `/dev/disk3s1`). Note that you *must* use

the command line or Disk Utility to unmount. If you just eject it from the Finder you'll have to take it out and reinsert it (and you'll still need to unmount it from the command line or Disk Utility). If the card fails to unmount, make sure to close any Finder windows that might be open on the card.

6. Next you'll need to figure out the raw device name of the card. Take your device name and replace `disk` with `rdisk` and leave off the s1 (which is the partition number). For example, the raw device name for the device `/dev/disk3s1` is `/dev/rdisk3`.

 It is really important that you get the raw device name correct! You can overwrite your hard drive and lose data if you start writing to your hard drive instead of the SD card. Use `df` again to double check before you continue.

---

7. Make sure that the downloaded image is unzipped and sitting in your home directory. You'll be using the Unix utility `dd` to copy the image bit by bit to the SD card. Below is the command; just replace the name of the disk image with the one you downloaded, and replace `/dev/rdisk3` with the raw device name of the SD card from step 6.

You can learn more about the command line in Chapter 3, but you're essentially telling `dd` to run as root and copy the input file (`if`) to the output file (`of`).

```
sudo dd bs=1m if=~/2012-09-18-wheezy-raspbian.img of=/dev/rdisk3
```

8. It will take a few minutes to copy the whole disk image. Unfortunately `dd` does not provide any visual feedback, so you'll just have to wait. When it's done it will show you some statistics; eject the SD card and you're ready to try it on the Pi.

Writing an SD card from Windows

1. Download the Win32DiskImager (*https://launchpad.net/win32-image-writer*) program.

2. Insert the SD card in your reader and note the drive letter that pops up in Windows Explorer.

3. Open Win32DiskImager and select the Raspbian disk image.

4. Select the SD card's drive letter, then click Write. If Win32DiskImager has problems writing to the card, try reformatting it in Windows Explorer.

5. Eject the SD card and put it in your Raspberry Pi; you're good to go!

Writing an SD card from Linux

The instructions for Linux are similar to those for the Mac:

1. Open your a new shell and *without* the card in the reader, type `df -h` to see which disk volumes are mounted.

2. Now insert the SD card and run `df -h` again.

3. Look at the list of mounted volumes and determine which one is the SD card by comparing it to the previous output. Find the device name, which should be something like `/dev/sdd1`. Depending on the configuration of your computer, this name may vary. Write the card's device name down.

4. To write to the card you'll have to unmount it first. Unmount it by typing `umount /dev/sdd1` (using the device name you got from the previous step instead of `/dev/sdd1`). If the card fails to unmount, make sure it is not the current working directory in any open shells.

5. Next you'll need to figure out the raw device name of the card, which is the device name without the partition number. For example, if your device name was `/dev/sdd1`, the raw device name is `/dev/sdd`.

 It is really important that you get the raw device name correct! You can overwrite your hard drive and lose data if you start writing to your hard drive instead of the SD card. Use `df` again to double check before you continue.

---

7. Make sure that the downloaded image is unzipped and sitting in your home directory. You'll be using the Unix utility `dd` to copy the image bit by bit to the SD card. Below is the command; just replace the name of the disk image with the one you downloaded, and replace `/dev/rdisk3` with the raw device name of the SD card from step 6.

```
sudo dd bs=1M if=~/2012-09-18-wheezy-raspbian.img of=/dev/sdd
```

This command tells `dd` to run as root and copy the input file (`if`) to the output file (`of`).

8. It will take a few minutes to copy the whole disk image. Unfortunately `dd` does not provide any visual feedback, so you'll just have to wait.

When it's done it will show you some statistics. It should be ok to eject the disk, but just to make sure it is safe, run `sudo sync`, which will flush the filesystem write buffers.

9. Eject the card and insert it in your Raspberry Pi. Good to go!

BerryBoot

A second way to get the OS on to an SD Card under is to use the BerryBoot utility (*http://www.berryterminal.com/doku.php/berryboot*). BerryBoot is part of the BerryTerminal thin client project, and will let you put multiple operating systems on a single card. You put the BerryBoot image on an SD card, boot it up on the Raspberry Pi and an interactive installer allows you to select an OS from a list. Note that you'll have to be connected to a network for BerryBoot to work.

B/Astral Trespassers Complete

This appendix contains the complete Scratch scripts for all the sprites in the Astral Tresspassers game described in Chapter 5. The examples are complete in that chapter as well, but since Scratch is a visual programming language it may be helpful to see them all together in one place. The costume tabs for each sprite are also highlighted.

Figure B-1. *The sprite list and the five sprites.*

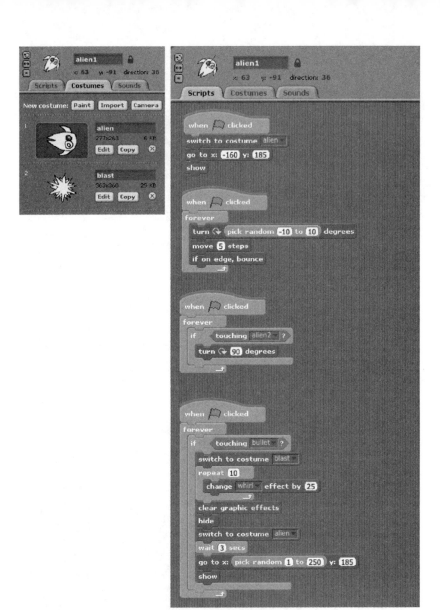

Figure B-2. *The alien1 sprite's costume tab (left) and complete script (right).*

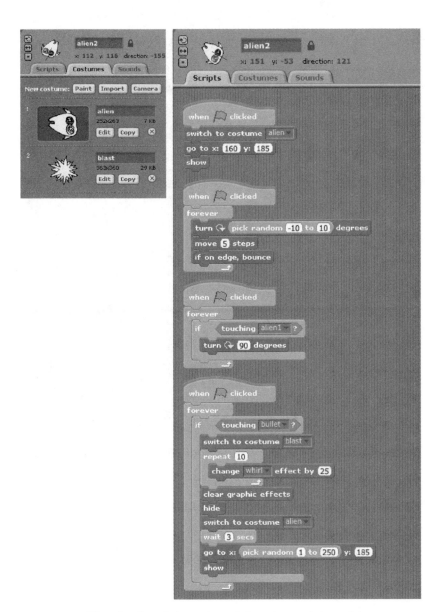

Figure B-3. *The alien2 sprite's costume tab and complete script.*

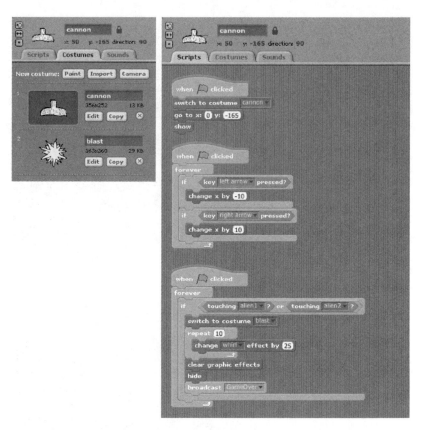

Figure B-4. *The cannon sprite's costume tab and complete script.*

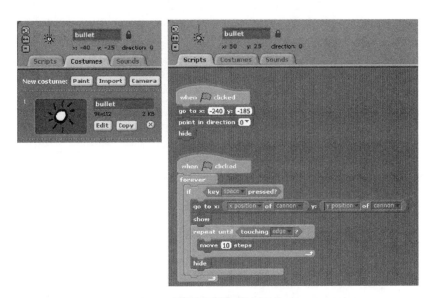

Figure B-5. *The bullet sprite's costume tab and complete script.*

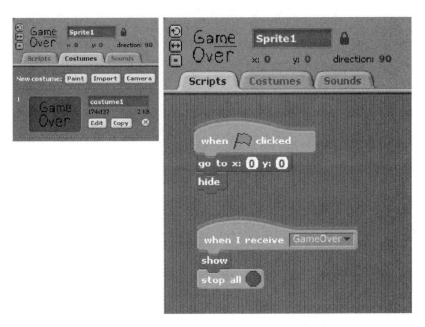

Figure B-6. *The gameover sprite's costume tab and complete script.*

C/Analog Input

In this book, you learned about digital inputs and outputs with buttons, switches, LEDs and relays. Each of these components were always either on or off, never anything in between. However, you might want to sense things in the world that are not necessarily one or the other, for instance: temperature, distance, light levels, or the status of a dial.

Unfortunately, you can't read these types of sensors directly on the Raspberry Pi since there are only digital inputs on the board. But you're not totally out of luck; there are a few different ways of using additional components to read analog sensors using the Pi's digital inputs.

Converting Analog to Digital

To convert from analog to digital, this appendix will show you how to use an *ADC*, or *analog to digital converter*. There are a few different models of ADC's out there, but we'll be using the ADS1015 from Texas Instruments. The package of the ADS1015 chip is too small for your standard breadboard, so Adafruit Industries has created a breakout board for it (*http://www.adafruit.com/products/1083*), shown in Figure C-1. Once you've soldered header pins to the breakout board, you can prototype with this chip in your breadboard. The chip uses a protocol called I2C for transmitting the analog readings. Luckily, we don't need to understand the protocol in order to use it. Adafruit provides an excellent open source Python library to read the values from the ADS1015 and its big brother, the ADS1115, via I2C.

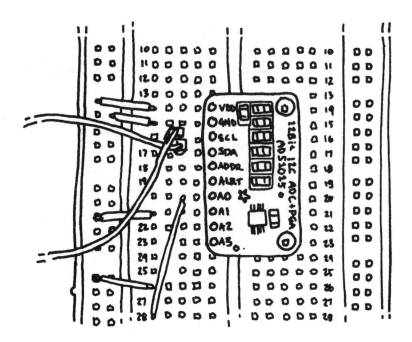

Figure C-1. *The ADS1015 analog to digital converter breakout from Ada-fruit*

To connect the ADS1015 breakout to your Raspberry Pi:

1. Connect the 3.3 volt pin from the Raspberry Pi to the positive rail of the breadboard. Refer to Figure 7-2 for pin locations on the Raspberry Pi's GPIO header.

2. Connect the ground pin from the Raspberry Pi to the negative rail of the breadboard.

3. Insert the ADS1015 into the breadboard and use jumper wires to connect its VDD pin to the positive rail and its GND pin to the negative rail.

4. Connect the SCL pin on the ADS1015 to the SCL pin on the Raspberry Pi. The SCL pin on the Pi is the one paired with the ground pin on the GPIO header.

5. Connect the SDA pin on the ADS1015 to the SDA pin on the Raspberry Pi. The SDA pin is in between the the SCL pin and the 3.3 volt pin.

Now you'll need to connect an analog sensor to the ADS1015. There are many to choose from, but for this walk-through, we'll use a 2K potentiometer so that we can have a dial input for our Raspberry Pi. A *potentiometer*, or *pot*, is essentially a variable resistor and can come in the form of a dial or slider. To connect a potentiometer to the ADS1015:

1. Insert the potentiometer into the breadboard.
2. The pot has three pins. Connect the middle pin to pin A0 in on the ADS1015.
3. One of the outside pins should connect to the positive rail of the bread-board. For now, it doesn't matter which.
4. Connect the other outside pin to the negative rail of the breadboard.

The connections should look as shown in Figure C-2.

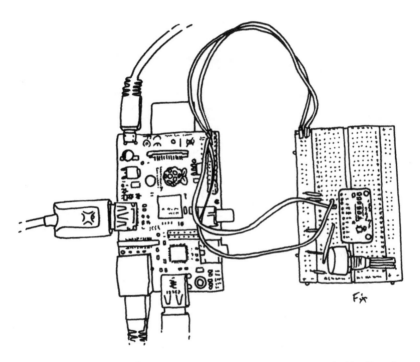

Figure C-2. *Using the ADS1015 to connect a potentiometer to the Raspberry Pi*

Before you can read the potentiometer, you'll need to enable I2C and install a couple libraries:

1. On the command line, open up the raspi-blacklist.conf file as root:

```
pi@raspberrypi ~ $ sudo nano /etc/modprobe.d/raspi-blacklist.conf
```

In order to remove I2C from this blacklist, put a hash mark in front of the line that says `blacklist i2c-bcm2708`. It should look like this:

```
# blacklist spi and i2c by default (many users don't need them)

blacklist spi-bcm2708
#blacklist i2c-bcm2708
```

2. Type `Control-X` to exit and then y to save the file.
3. Next, open /etc/modules:

```
pi@raspberrypi ~ $ sudo nano /etc/modules
```

4. Add `i2c-dev` to the end of the file, on its own line. The file should look like this:

```
# /etc/modules: kernel modules to load at boot time.
#
# This file contains the names of kernel modules that should be loaded
# at boot time, one per line. Lines beginning with "#" are ignored.
# Parameters can be specified after the module name.

snd-bcm2835
i2c-dev
```

5. Type `Control-X` to exit and then y to save the file.
6. Update your list of packages:

```
pi@raspberrypi ~ $ sudo apt-get update
```

7. Install `i2c-tools` tools and `python-smbus`:

```
pi@raspberrypi ~ $ sudo apt-get install i2c-tools python-smbus
```

8. Restart your Raspberry Pi.
9. After you've restarted your Raspberry Pi, test that the Raspberry Pi can detect the ADS1015. On revision 1 boards, use the command:

```
pi@raspberrypi ~ $ sudo i2cdetect -y 0
```

On revision 2 boards, use the command:

```
pi@raspberrypi ~ $ sudo i2cdetect -y 1
```

10. If the board is recognized, you'll see the number in the grid that is displayed:

```
     0  1  2  3  4  5  6  7  8  9  a  b  c  d  e  f
00:        -- -- -- -- -- -- -- -- -- -- -- -- -- --
10: -- -- -- -- -- -- -- -- -- -- -- -- -- -- -- --
20: -- -- -- -- -- -- -- -- -- -- -- -- -- -- -- --
30: -- -- -- -- -- -- -- -- -- -- -- -- -- -- -- --
40: -- -- -- -- -- -- -- -- 48 -- -- -- -- -- -- --
50: -- -- -- -- -- -- -- -- -- -- -- -- -- -- -- --
60: -- -- -- -- -- -- -- -- -- -- -- -- -- -- -- --
70: -- -- -- -- -- -- -- --
```

11. Now that we know that the device is connected and is recognized by our Pi, it's time to start reading the potentiometer. To do so, download the Raspberry Pi Python libraries from Adafruit's code repository into your home folder. Type the following command at the shell prompt, all on one line with no spaces in the URL:

```
wget    https://github.com/adafruit/Adafruit-Raspberry-Pi-Python-Code/
archive/master.zip
```

12. Unzip it:

```
pi@raspberrypi ~ $ unzip master.zip
```

13. Change to the library's ADS1x15 directory:

```
$ cd Adafruit-Raspberry-Pi-Python-Code-master/Adafruit_ADS1x15
```

14. Run the example file:

```
$ sudo python ads1015_example.py
Channel 0 = 2.067 V
Channel 1 = 3.309 V
```

15. Turn the potentiometer all the way in one direction and run it again. Notice the change in the value for channel 0:

```
$ sudo python ads1015_example.py
Channel 0 = 3.306 V
Channel 1 = 3.309 V
```

16. Turn the potentiometer all the way in the other direction and run it one more time:

```
$ sudo python ads1015_example.py
Channel 0 = 0.000 V
Channel 1 = 3.309 V
```

As you can see, turning the dial on the potentiometer changes the voltage coming into channel 0 of the ADS1015. The code in *ads1015_example.py* does a little bit of math to determine the voltage value from the data coming from the ADC. Of course, your math will vary depending on what kind of sensor you want use.

Try creating a new file in the same directory with the following code:

```
from Adafruit_ADS1x15 import ADS1x15 ❶
from time import sleep

adc = ADS1x15() ❷

while True:
    result = adc.readADCSingleEnded(0) ❸
    print result
    sleep(.5)
```

❶   Import Adafruit's ADS1x15 library

❷   Create a new ADS1x15 object called adc.

❸   Get a reading from channel A0 on the ADS1015 and store it in result.

When you run this code as root, it will output out raw numbers for each reading twice a second. Turning the potentiometer will make the values go up or down.

Once you get it all set up, the Adafruit ADS1x15 library does all the hard work for you and makes it easy to use analog sensors in your projects. For instance, if you want to make your own Pong-like game, you could read two potentiometers and then use Pygame to draw on the game on screen. For more information about using Pygame, see Chapter 4.

Adafruit's Educational Linux Distro

Adafruit Industries created a fork of the main Raspbian/Wheezy Linux distribution and added drivers and software that make electronics prototyping easier right out of the box. The result is the Adafruit Raspberry Pi Educational Linux Distro (*http://learn.adafruit.com/adafruit-raspberry-pi-educational-linux-distro*), codenamed *Occidentalis* from the species name of the black raspberry.

Occidentalis is an interesting example of the open source model in action; the Linux ecosystem has a long history of forked distributions that are specially tuned for a particular application. Often these forks are proving grounds for targeted improvements that can be merged back into other distributions.

Many of the improvements and additions in the distribution are aimed at supporting popular sensors and some Adafruit products without having to install additional software. There are also some low-level improvements that make it easier to perform Pulse Width Modulation and control servo motors directly from the Raspberry Pi.

See the overview (*http://learn.adafruit.com/adafruit-raspberry-pi-educational-linux-distro*) page for a list of features, and the Adafruit tutorials (*http://learn.adafruit.com/category/raspberry-pi*) for examples of how to use Occidentalis.

About the Authors

Matt Richardson is a Brooklyn-based creative technologist and video producer. He's a contributor to *MAKE* magazine and Makezine.com. Matt is also the owner of Awesome Button Studios, a technology consultancy. Highlights from his work include the Descriptive Camera, a camera which outputs a text description of a scene instead of a photo. He also created The Enough Already, a DIY celebrity-silencing device. Matt's work has garnered attention from *The New York Times*, *Wired*, *New York Magazine* and has also been featured at The Nevada Museum of Art and at the Santorini Bienniele. He is currently a Master's candidate at New York University's Interactive Telecommunications Program.

Shawn Wallace is an editor at O'Reilly and lives in Providence, RI. He is also a member of the Fluxama artist collective responsible for new iOS musical instruments such as Noisemusick and Doctor Om. He designed open hardware kits at Modern Device and taught the Fab Academy at the Providence Fab Lab. For years he was the managing director of the AS220 art space and is a cofounder of the SMT Computing Society.

The cover and body font is BentonSans, the heading font is Serifa, and the code font is Bitstreams Vera Sans Mono.

Have it your way.

Get even more for your money.

Join the O'Reilly Community, and register the O'Reilly books you own. It's free, and you'll get:

- $4.99 ebook upgrade offer
- 40% upgrade offer on O'Reilly print books
- Membership discounts on books and events
- Free lifetime updates to ebooks and videos
- Multiple ebook formats, DRM FREE
- Participation in the O'Reilly community
- Newsletters
- Account management
- 100% Satisfaction Guarantee

Signing up is easy:

1. Go to: oreilly.com/go/register
2. Create an O'Reilly login.
3. Provide your address.
4. Register your books.

Note: English-language books only

To order books online:

oreilly.com/store

For questions about products or an order:

orders@oreilly.com

To sign up to get topic-specific email announcements and/or news about upcoming books, conferences, special offers, and new technologies:

elists@oreilly.com

For technical questions about book content:

booktech@oreilly.com

To submit new book proposals to our editors:

proposals@oreilly.com

O'Reilly books are available in multiple DRM-free ebook formats. For more information:

oreilly.com/ebooks

O'REILLY®